Cellular Entry of Binary and Pore-Forming Bacterial Toxins

Special Issue Editor
Alexey S. Ladokhin

MDPI • Basel • Beijing • Wuhan • Barcelona • Belgrade

MDPI

Special Issue Editor
Alexey S. Ladokhin
The University of Kansas Medical Center
USA

Editorial Office
MDPI AG
St. Alban-Anlage 66
Basel, Switzerland

This edition is a reprint of the Special Issue published online in the open access journal *Toxins* (ISSN 2072-6651) from 2017–2018 (available at: http://www.mdpi.com/journal/toxins/special issues/pore forming bacterial toxins).

For citation purposes, cite each article independently as indicated on the article page online and as indicated below:

Lastname, F.M.; Lastname, F.M. Article title. *Journal Name*. **Year**. *Article number*, page range.

First Edition 2018

Image courtesy of Alexandra J. Machen and Mark T. Fisher

ISBN 978-3-03842-704-9 (Pbk)
ISBN 978-3-03842-703-2 (PDF)

Table of Contents

About the Special Issue Editor

Alexey S. Ladokhin was born in Donetsk (Донецьк), Ukraine and got his undergraduate degree in Physics from Shevchenko National University in Kyiv (Київ) in 1984. He received his Ph.D. degree in Biophysics and D.Sc. degree in Molecular Biology from the National Academy of Sciences of Ukraine. Dr. Ladokhin had worked as a research associate at the University of Virginia, Johns Hopkins Univer-sity and the University of California at Irvine, before assuming a faculty position at the Department of Biochemistry and Molecular Biology of the University of Kansas Medical Center in 2004. His re-search interests include folding and insertion of membrane proteins; membrane action of bacterial toxins, apoptotic regulators, antimicrobial and toxic peptides; thermodynamics of membrane protein insertion and assembly; development of fluorescence methods for membrane studies; integration of experimental and computational methods.

toxins

MDPI

Editorial

Cellular Entry of Binary and Pore-Forming Bacterial Toxins

Alexey S. Ladokhin

Department of Biochemistry and Molecular Biology, University of Kansas Medical Center, Kansas City, KS 66160, USA; aladokhin@kumc.edu; Tel.: +1-913-588-0489

Received: 21 December 2017; Accepted: 22 December 2017; Published: 26 December 2017

This Special Issue of *Toxins*, entitled "Cellular Entry of Binary and Pore-Forming Bacterial Toxins," gives a sense of the recent advances in characterizing the functional and structural aspects of this broad scientific problem that goes beyond the classical field of toxinology and microbiology and spills into the general areas of biochemistry, biophysics, and molecular and cell biology. The contributions to this Special Issue include several experimental articles, employing sophisticated techniques to gain important insights into the mechanism of cellular entry [1–6]; a thought-provoking perspective comment [7]; and two conceptual reviews, one on apicomplexan pore-forming toxins [8] and one on clostridial binary toxins [9]. What have we learned about the field from this collection? Despite the limited selection, some general features can be identified.

Deciphering complex pathways requires integration of various approaches. Cellular entry of bacterial toxins utilizes a complex mechanism [8,9] that involves multiple protein partners interacting with each other [3,9] and with a lipid bilayer [1,2,6]. Key players often undergo profound conformational changes, both in aqueous [5] and membranous environments [1,2]. Characterizing these functionally important conformational changes is a prerequisite for deciphering the mechanisms of cellular entry on a molecular level. One of the biggest challenges in establishing the structure–function relationships for bacterial toxins lies in their environment-dependent conformational lability. Consequently, even if a high-resolution structure of the soluble conformation is well-characterized, the mechanism might remain elusive, due to conformational rearrangements triggered by environment acidification and membrane insertion, common for the endosome-dependent pathways. These challenges could be met, for example, by careful examination of site-directed mutagenesis with a variety of functional assays (e.g., for diphtheria toxin [6]), complemented with molecular modeling (e.g., for perfringolysin O [1]). In another example, a sophisticated combination of cryo-electron microscopy, performed on elaborately prepared nanodisc samples, and computer simulations is used to resolve the structure of the pore of the anthrax toxin protective antigen in a lipid environment and in a complex with the toxin's lethal factor [2].

Structured vs. unstructured passageways through the membrane. Bridging cellular membranes is a key step in the pathogenic action of both binary and pore-forming toxins. The former use their translocation domains, containing various structural motifs, to ensure efficient delivery of the toxic component into the host cell, while the latter act on the cellular membrane itself. In either case, the integrity of the membrane is compromised via targeted protein–lipid and protein–protein interactions triggered by specific signals, such as proteolytic cleavage and/or endosomal acidification. Several studies presented in this Special Issue either explicitly describe the formation of the water-filled protein structures that span the lipid-bilayer or implicitly evoke such structures, as a required part of the cellular entry mechanism. Specific structural examples that include both binary (e.g., anthrax [2]) and pore-forming toxins (e.g., perfringolysin O [1]) involve the insertion of the β-strands from multiple protein subunits to form a barrel-like structure that bridges the lipid bilayer in a permanent way. A similar concept has been evoked for other toxins as well, the translocation domains of which form α-helices in the lipid bilayer. Specifically, the translocation domain of diphtheria toxin was often assumed to use the so-called open-channel state (OCS), formed by three transmembrane helices, as a translocation pathway. The examination of both in vivo and in vitro activity of the several OCS-blocking mutants, presented in this issue [6], revealed that the OCS is formed after the translocation, which is likely to utilize an unstructured and possibly transient passageway. Certainly, more studies with other toxins are needed before any general conclusions can be reached on the possible differences between the actions of toxins that utilize α-helical vs. β-structure motifs in their membrane-interacting domains. More studies are also needed to fully characterize the structural and thermodynamic aspects of the conformational switching and membrane interactions involved in the cellular entry of bacterial protein toxins. Deciphering the physicochemical principles underlying these processes is also a prerequisite for the use of protein engineering to develop toxin-based molecular vehicles capable of targeted delivery of therapeutic agents to tumors and other diseased tissues.

Conflicts of Interest: The author declares no conflict of interest.

References

1. Savinov, S.N.; Heuck, A.P. Interaction of Cholesterol with Perfringolysin O: What Have We Learned from Functional Analysis? *Toxins* **2017**, *9*, 381. [CrossRef] [PubMed]
2. Machen, A.J.; Akkaladevi, N.; Trecazzi, C.; O'Neil, P.T.; Mukherjee, S.; Qi, Y.; Dillard, R.; Im, W.; Gogol, E.P.; White, T.A.; et al. Asymmetric Cryo-EM Structure of Anthrax Toxin Protective Antigen Pore with Lethal Factor N-Terminal Domain. *Toxins* **2017**, *9*, 298. [CrossRef] [PubMed]
3. Tausch, F.; Dietrich, R.; Schauer, K.; Janowski, R.; Niessing, D.; Martlbauer, E.; Jessberger, N. Evidence for Complex Formation of the *Bacillus cereus* Haemolysin BL Components in Solution. *Toxins* **2017**, *9*, 288. [CrossRef] [PubMed]
4. Puri, M.; La Pietra, L.; Mraheil, M.A.; Lucas, R.; Chakraborty, T.; Pillich, H. Listeriolysin O Regulates the Expression of Optineurin, an Autophagy Adaptor That Inhibits the Growth of *Listeria monocytogenes*. *Toxins* **2017**, *9*, 273. [CrossRef] [PubMed]
5. Palma, L.; Scott, D.J.; Harris, G.; Din, S.U.; Williams, T.L.; Roberts, O.J.; Young, M.T.; Caballero, P.; Berry, C. The Vip3Ag4 Insecticidal Protoxin from *Bacillus thuringiensis* Adopts A Tetrameric Configuration That Is Maintained on Proteolysis. *Toxins* **2017**, *9*, 165. [CrossRef] [PubMed]
6. Ladokhin, A.S.; Vargas-Uribe, M.; Rodnin, M.V.; Ghatak, C.; Sharma, O. Cellular Entry of the Diphtheria Toxin Does Not Require the Formation of the Open-Channel State by Its Translocation Domain. *Toxins* **2017**, *9*, 299. [CrossRef] [PubMed]
7. Knap, P.; Tebaldi, T.; Di Leva, F.; Biagioli, M.; Dalla Serra, M.; Viero, G. The Unexpected Tuners: Are LncRNAs Regulating Host Translation during Infections? *Toxins* **2017**, *9*, 357. [CrossRef]
8. Guerra, A.J.; Carruthers, V.B. Structural Features of Apicomplexan Pore-Forming Proteins and Their Roles in Parasite Cell Traversal and Egress. *Toxins* **2017**, *9*, 265. [CrossRef]
9. Takehara, M.; Takagishi, T.; Seike, S.; Oda, M.; Sakaguchi, Y.; Hisatsune, J.; Ochi, S.; Kobayashi, K.; Nagahama, M. Cellular Entry of *Clostridium perfringens* Iota-Toxin and *Clostridium botulinum* C2 Toxin. *Toxins* **2017**, *9*, 247. [CrossRef] [PubMed]

toxins

MDPI

Review

Cellular Entry of *Clostridium perfringens* Iota-Toxin and *Clostridium botulinum* C2 Toxin

Masaya Takehara [1], Teruhisa Takagishi [1], Soshi Seike [2], Masataka Oda [3], Yoshihiko Sakaguchi [4], Junzo Hisatsune [5], Sadayuki Ochi [6], Keiko Kobayashi [1] and Masahiro Nagahama [1,*]

[1]   Department of Microbiology, Faculty of Pharmaceutical Sciences, Tokushima Bunri University, Yamashiro-cho, Tokushima 770-8514, Japan; mtakehara@ph.bunri-u.ac.jp (M.T.); t.takagishi@ph.bunri-u.ac.jp (T.T.); kobakei@ph.bunri-u.ac.jp (K.K.)

[2]   Laboratory of Molecular Microbiological Science, Faculty of Pharmaceutical Sciences, Hiroshima International University, Kure, Hiroshima 737-0112, Japan; s-seike@ps.hirokoku-u.ac.jp

[3]   Department of Microbiology and Infection Control Science, Kyoto Pharmaceutical University, Yamashina, Kyoto 607-8414, Japan; moda@mb.kyoto-phu.ac.jp

[4]   Department of Microbiology, Kitasato University School of Medicine, 1-15-1 Kitasato, Minami-ku, Sagamihara, Kanagawa 252-0374, Japan; ysakaguchi@med.kitasato-u.ac.jp

[5]   Department of Bacteriology, Graduate school of Biomedical and Health Sciences, Hiroshima University, 1-2-3 Kasumi, Minami-ku, Hiroshima 734-8551, Japan; hisatune@hiroshima-u.ac.jp

[6]   Faculty of Pharmacy, Yokohama University of Pharmacy, 601 Matano-cho, Totsuka-ku, Yokohama-shi, Kanagawa 245-0066, Japan; sadayuki.ochi@hamayaku.ac.jp

*   Correspondence: nagahama@ph.bunri-u.ac.jp; Tel.: +81-088-622-9611; Fax: +81-088-655-3051

Academic Editor: Alexey S. Ladokhin

Received: 19 July 2017; Accepted: 9 August 2017; Published: 11 August 2017

Abstract: *Clostridium perfringens* iota-toxin and *Clostridium botulinum* C2 toxin are composed of two non-linked proteins, one being the enzymatic component and the other being the binding/translocation component. These latter components recognize specific receptors and oligomerize in plasma membrane lipid-rafts, mediating the uptake of the enzymatic component into the cytosol. Enzymatic components induce actin cytoskeleton disorganization through the ADP-ribosylation of actin and are responsible for cell rounding and death. This review focuses upon the recent advances in cellular internalization of clostridial binary toxins.

Keywords: clostridial binary toxin; iota-toxin; C2 toxin; cellular internalization

1. Introduction

Clostridial binary toxins are ADP-ribosylating toxins that utilize globular actin as a substrate and depolymerize filamentous actin capped by ADP-ribosylated actin in sensitive cells. Clostridial binary toxins are produced by a few clostridia and are categorized into two groups [1–3]. One group consists of *Clostridium (C.) perfringens* (type E) iota-toxin [4], *C. difficile* transferase (CDT) [5], and *C. spiroforme* iota-like toxin [6]. The other group includes C2 toxin produced by *C. botulinum* types C and D [7]. The amino acid sequences of the former three binding components are more similar to each other than to C2II. These clostridial binary toxins consist of two separate protein components: the enzymatic A component and the binding/translocation B component. The binding components Ib and C2II (B components of iota-toxin and C2 toxin, respectively) specifically recognize different cellular receptors and are implicated in the uptake of A components into the intracellular space [3,8–11]. The A component (Ia) of iota-toxin mono-ADP-ribosylates non-muscle and muscle G-actin at arginine-177. On the other hand, the A component (C2I) of C2 toxin mono-ADP-ribosylates non-muscle G-actin [12]. These binary toxins cause the depolymerization of actin filaments and sensitive cells round-up as a result. Ib internalizes the A components from either iota-like toxin or CDT, but not

C2II [10,11,13]. Recent advances in our understanding of the cellular uptake of iota-toxin and C2 toxin provide fascinating insights into the mechanism of cytotoxicity.

2. *C. perfringens* Iota-Toxin

Iota-toxin produced by *C. perfringens* type E consists of two components, an enzymatic component (Ia) and a binding component (Ib) [2,3,10,11]. Each individual component is deficient in toxic activity, but the combination of Ia and Ib causes lethal, dermonecrotic, and cytotoxic activities. Type E strain infection leads to antibiotic-associated enterotoxemia in rabbits [14,15]. Moreover, type E strains have been associated with hemorrhagic enterocolitis and sudden death in calves and lambs [14,15]. Iota-toxin is considered to be a key virulence factor of intestinal pathogenesis.

2.1. Structure of Ia and Ib

Crystal structure analysis of Ia indicated that it is separated into two different domains: an N-domain, which plays a role in binding with Ib, and a C-domain, which is responsible for NAD binding and ADP-ribosylating activity [16]. Previously, we reported the crystal structure of a complex consisting of Ia, actin monomer, and a hydrolysis-resistant NAD$^+$ derivative [17]. On the basis of the structure of this complex, Tyr-62 and Arg-248 in Ia were shown to be critical for the Ia/actin interaction. A few conformational "snapshots" were identified, indicating that the formation of the Ia/actin complex formation occurs as part of the ADP-ribosyltransferase-catalyzed reaction. In addition, critical catalytic residues of Ia and of actin were identified. The structures confirmed a "strain-alleviation model" of ADP-ribosylation [18]. This finding suggested that all ADP-ribosyltransferases, including mono- and poly-ADP-ribosyltransferases, share a common catalytic mechanism.

Ib is produced as an inactive form (100 kDa). The active form of Ib (80 kDa) is generated by the proteolytic removal of a 20 kDa N-terminal fragment from the inactive form [1]. Ib shares 39% similarity overall with C2II (binding component of C2 toxin) [19]. Ib contains four distinct domains [2,3,10,11,19]. Domain 1 (N-terminal domain) provides the binding site for Ia; domain 4 (C-terminal domain) is a potential binding site for the host cell receptor; and domains 2 and 3 are respectively involved in oligomer assembly and pore formation. Ib assembles into heptamers, in a ring-shaped structure, which play a role in the translocation of Ia into the cytoplasm after internalization [10,11,19].

2.2. Binding and Internalization of Iota-Toxin

It has been reported that lipolysis-stimulated lipoprotein receptor (LSR) is a host cellular receptor for Ib [20,21], with iota-toxin entering host cells via an LSR-mediated process. Recently, we reported that domain 4 of Ib (Ib442-664) binds to LSR in a tricellular tight junction (tTJ) [22]. This binding led to the removal of LSR from the tTJ, which enhanced the permeation of macromolecular solutes, indicating that Ib442-664 is a modification factor of the tTJ barrier. This confirmed that domain 4 of Ib works as a receptor recognition site [2,3,10,19]. In contrast, it has also been shown that iota-toxin enters the host cells via cell-surface antigen CD44-dependent endocytosis [23].

Lipid raft membrane microdomains have been shown to serve as cell surface platforms for clustering bacteria, viruses, and several toxins [24–26]. These ligands have been reported to invade host cells through binding to lipid rafts. In addition, Ib oligomer clustered in plasma membrane lipid rafts as well as Ia associated with Ib oligomer invade the host cells [9,27]. Ib monomer is observed in lipid rafts and non-lipid raft fractions in whole plasma membrane at 37 °C, revealing that the Ib receptor is distributed throughout the plasma membranes. Thus, the receptor is not constrained to membrane lipid rafts. Because Ib oligomer is observed in membrane lipid rafts after treatment at 37 °C, the binding of Ib to the receptor causes movement from non-lipid rafts to lipid rafts, resulting in the formation of oligomers in the microdomains. Ib421-664 inhibited Ib binding to the target cells [9,28], and was localized in membrane lipid rafts. On the basis of these findings, because domain 4 of Ib associates with the receptor that is distributed throughout the whole cytoplasmic membrane, the receptor, which is bound to Ib, moves into membrane lipid rafts [9]. Papatherodorou et al. [20]

reported that the binding component causes the accumulation of LSR in membrane lipid rafts, and LSR accumulation is a potent trigger for the oligomerization of Ib. Moreover, CD44 is mainly observed in membrane lipid rafts obtained from host cells incubated with iota-toxin [29], and CD44 promotes the accumulation of LSR into membrane lipid rafts [30].

2.3. Intracellular Trafficking of Iota-Toxin

C. perfringens iota-toxin internalizes in sensitive cells and causes cytotoxicity by utilizing the endocytic pathway [2,3,10,11,19]. Ib interacts directly with a single cellular receptor (e.g., LSR), promotes oligomerization on membrane lipid rafts, and associates with Ia [9,13,20,27]. After internalization via a Rho-independent and clathrin-dependent pathway, the toxin moves through the pathway until it reaches endocytic vesicles [13,31]. Following a 15-min incubation of Ib with host cells at 37 °C, it is detected in early endosomes (EEs) [32]. However, after 30 min, Ib is not observed in EEs and, 15–30 min later, low levels of Ib are transported to Rab11-positive recycling endosomes (REs). Therefore, a small proportion of Ib is driven back to the plasma membranes by the salvage mechanism for intracellular recycling via REs. This recycling process is critical for Ib to augment the uptake of Ia. After 30–60 min, Ib is delivered to late endosomes (LEs) and lysosomes. On the basis of these findings, Ib is internalized and delivered from EEs to REs, or LEs and lysosomes. Lysosomes migrate to cell membranes through a Ca^{2+}-mediated mechanism and fuse with cell membranes [32]. Because Ib causes the elevation of intracellular Ca^{2+} from the extracellular medium, Ib induces fusion between lysosomes and cell membranes. This fusion promotes the repair of damaged membranes during Ib pore formation (Figure 1).

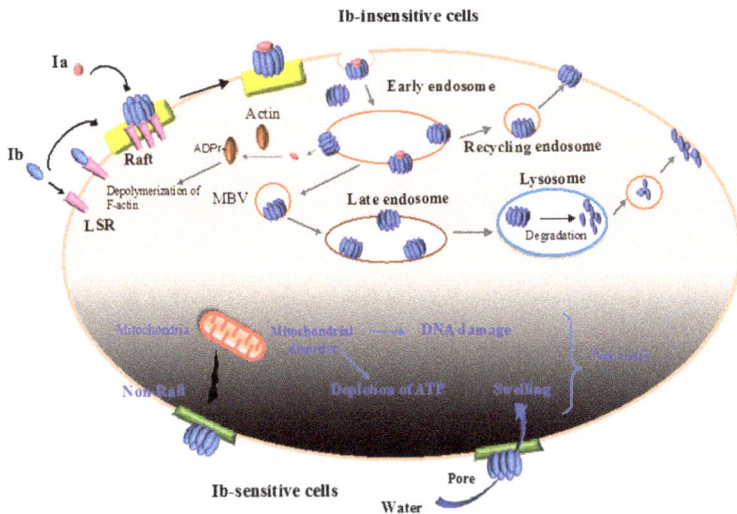

Figure 1. Mode of action of iota-toxin in various cells. Upper part (Ib-insensitive cells): Ib associates with a receptor (LSR: lipolysis-stimulated lipoprotein receptor) on the plasma membrane and migrates to membrane lipid raft; Ia bound to Ib oligomers forms on the rafts. Then, the Ia and Ib complex enters the cell. The complex is trafficked to the early endosome, where acidification facilitates the cytosolic release of Ia. Ia ADP-ribosylates G-actin in the cytoplasm, ultimately causing cytotoxicity. Ib is sorted into recycling endosomes and late endosomes. From recycling endosomes, Ib is sent back to the plasma membranes, and this recycling process is critical for Ib to enhance the entry of Ia. From late endosomes, Ib is delivered to lysosomes for degradation, and degraded Ib is exposed on the cell surface. Lower part (Ib-sensitive cells): Ib oligomerizes mainly in non-lipid rafts in the plasma membranes and is not internalized. Ib induces mitochondrial damage, and subsequently gives rise to the depletion of ATP and DNA damage. On the other hand, Ib causes the swelling of cells mediated by Ib pores. Finally, Ib induces cell necrosis.

2.4. Translocation of Ia across the Endosomal Membrane

After the endocytosis of iota-toxin, Ia passes through the endosomal membrane into the cytoplasm. This process uses the endosomal membrane-spanning pore formed by Ib. Acidic environments are essential for the translocation of Ia from EEs into the cytoplasm. Acidic pH causes structural alterations in the Ib oligomer, accelerating the insertion of the Ib oligomer into endosomal membranes and in turn the migration of Ia via oligomeric Ib pores into the cytoplasm after the pH gradient [31]. Ia is partly unfolded in order to migrate via the narrow oligomeric Ib pore into the cytoplasm. The pH-dependent transmembrane transport or cytoplasmic refolding of Ia is promoted by cytoplasmic factors containing the molecular chaperone heat shock protein 90 (Hsp90) and peptidyl-prolyl *cis/trans*-isomerase (PPIase), including cyclophilin A and FK-506 binding protein [33]. PPIase is a folding helper protein. Suppression of Hsp90 and PPIase blocks the transmembrane transport of Ia into the cytoplasm, and Ia binds to Hsp90 and PPIase in dot-blot analyses. Recently, it has been reported that an Hsp70 inhibitor blocks cytotoxicity induced by iota-toxin [34]. Hsp70 assists protein transport during transit through membranes. Hsp70, Hsp90, and PPIase cooperatively configure multi-chaperone complexes critical for the protein folding, intracellular localization, and maturation of particular proteins. The unfolded Ia has been shown to undergo potent binding with Hsp70 and PPIase compared with the native form. Together, Hsp70, Hsp-90, and PPIase are important for transport of Ia through the endosomal membrane [33–35].

2.5. Cytotoxicity of Ib

Ib alone has been shown to lack toxic activity [2,3,10,11,19]. We have reported that, after the binding of Ib to the cell surface receptor on Vero cells, it forms oligomers and creates ion-permeable pores [8]. The formation of pores by Ib has been demonstrated in planar lipid bilayers [36]. Ib creates the ion-permeable channels [36], and domains 2 and 3 in Ib are critical for oligomerization and the formation of channels, respectively [36,37]. Moreover, Ib leads to a decrease in transepithelial electrical resistance (TEER) in human intestinal epithelial Caco-2 cell monolayers [37]. We demonstrated that Ib alone has cytotoxic activity, and we examined the effects of Ib alone in eight cell lines [38]. Ib rapidly caused cell swelling, depletion of ATP, and reduction in viability among human epithelial carcinoma A431 and human lung adenocarcinoma A549 cells [38]. In MDCK cells, which are not sensitive to the cytotoxic activity of Ib, the Ib oligomer formed in membrane lipid rafts is taken up by endocytosis [9,38]. However, Ib also formed oligomers in non-raft membranes in A431 cells [38]. Additionally, Ib was present on the A431 cell surface while exhibiting its toxic activity. Long-term persistence of Ib in cell surfaces was dependent on the cell type, and internalization of Ib was linked with the survival of the challenge of Ib pores (Figure 1). Therefore, the ability of a cell type to survive membrane perforation by Ib depends on its ability to internalize Ib. Our results showed that the endocytosis of Ib is needed for host cell survival, and the function of endocytosis is an innate host defense response against pore-forming proteins [38].

Because no cell line was known to be susceptible to Ib until now, the finding that Ib induces cell death in A431 and A549 cells is a new discovery, and this may help to elucidate the contribution of Ib to the virulence of type E strains.

3. *C. botulinum* C2 Toxin

C2 toxin produced by *C. botulinum* type C and D is composed of an enzymatic subunit (C2I) and a binding/translocation subunit (C2II). Each protein component is generated separately and they are not linked. *C. botulinum* C2 toxin occasionally causes enteric hemorrhagic and necrotizing damage in animals, especially in avian species, which die of poisoning due to *C. botulinum* [11]. Experimentally, C2 toxin induces dermonecrosis and hemorrhagic enterocolitis in mice [11].

3.1. Structure of C2I and C2II

From its crystal structure, C2I is composed of two nearly equal-sized domains of about 200 residues [39]. Residues 1 to 87 of the N-terminal domain serve as a binding site for C2II. The C-terminal domain of C2I is involved in actin ADP-ribosyltransferase activity [2,10,19,39]. The C-terminal domain harbors highly conserved catalytic residues amongst bacterial ADP-ribosyltransferases.

C2II has to be activated by proteolytically cleaving a ~20 kDa N-terminal fragment (C2IIa) [2,10,11]. The structure of C2II has been determined and indicates that its structure is similar to that of protective antigen (PA), the binding component of *Bacillus anthracis* [39]. C2II exhibits four functional domains similar to iota-toxin [2,10,11]. The N-terminal domain (domain 1) includes the docking region with C2I; domain 2 is essential for the oligomer formation; the biological function of domain 3 is unclear; and the C-terminal domain (domain 4) plays a role in the recognition of the host cell receptor.

3.2. Binding and Internalization of C2 Toxin

Domain 4 of C2II has been determined as the binding domain for the host cell receptor [2,10,11]. C2II oligomers bind to asparagine-linked carbohydrates at the cell surface receptor [40]. The low level of amino-acid identity (lower than approximately 10%) between the sequence of these binding domains agrees well with the fact that each one binds to different receptors.

Recently, we demonstrated that C2 toxin needs the activity of acid sphingomyelinase (ASMase) during the initial step of endocytosis [41]. Several bacterial pore-forming cytotoxic proteins cause calcium entry and provoke the exocytosis of lysosomes, leading to the release of lysosomal ASMase to the extracellular medium [42,43]. Next, ASMase converts sphingomyelin in the outer leaflet of the plasma membrane to ceramide [44]. Then, ceramide self-assembles into ceramide-enriched microdomains that bud into cytoplasmic membranes, creating endosomes [45]. Namely, the hydrolysis of sphingomyelin by ASMase secreted from the exocytosis of lysosomes produces plasma membrane microdomains enriched in ceramide, leading to endocytosis [44,45]. C2IIa causes the entry of extracellular calcium into sensitive cells, and the cytotoxic activity of C2 toxin is increased in calcium-containing medium [41]. Blockers of lysosomal exocytosis and ASMase inhibit the cytotoxic activity of C2 toxin. Moreover, C2IIa induces the release of ASMase due to the exocytosis of lysosomes. Then, C2 toxin induces ceramide production in plasma membranes. On the basis of these findings, it is concluded that ASMase activity is required for C2 toxin entry into host cells [41]. In Figure 2, we show a hallmark of the key role of ASMase in the endocytosis of C2 toxin in sensitive cells.

C2IIa oligomer binds to plasma membrane lipid rafts [46]. By surface plasmon resonance analysis, it has been shown that C2I associates with oligomers of C2IIa but not with C2IIa monomers [46]. The binding of C2I to lipid raft-associated C2IIa oligomers induces rapid internalization of C2 toxin into host cells. The entry of C2 toxin proceeds through membrane lipid rafts, indicating that these structures include fundamental factors facilitating the internalization of C2 toxin. Hence, the C2I-C2IIa complex is endocytosed via plasma membrane lipid rafts [46].

Phosphatidylinositol 3-kinase (PI3K) and Akt inhibitors blocked the endocytosis of C2 toxin in the host cells and the cytotoxic effect of C2 toxin [46]. In fact, C2 toxin induced the activation of PI3K and the phosphorylation of Akt [46]. As mentioned above, C2 toxin causes the production of ceramide as a result of the ASMase activity [41]. Ceramide is metabolized to ceramide-1-phosphate (C1P). As stimulation of a C1P receptor is known to activate the PI3K/Akt signaling pathway, activation of this pathway by C2 toxin is needed for the production of C1P [46]. On the other hand, activation of the PI3K/Akt signaling pathway promotes cell survival [47,48]. Accordingly, antagonistic modes of action on the cytotoxicity act as the cellular defense mechanism against internalizing C2 toxin.

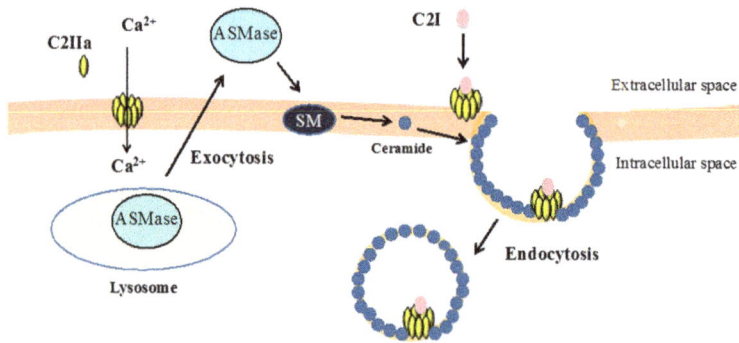

Figure 2. Initial step in the internalization of C2 toxin. Extracellular Ca^{2+} entry into the cytosol via a C2IIa pore. An increase in the intracellular Ca^{2+} concentration evokes lysosomal exocytosis. Lysosomal acid sphingomyelinase (ASMase) is secreted to the outer plasma membrane, where it hydrolyzes sphingomyelin into ceramide. Ceramide self-assembles into microdomains that bend to the intracellular space, elaborating endosomes that endocytose the C2I and C2IIa complex.

3.3. Intracellular Trafficking of C2 Toxin

The association of C2I with C2IIa oligomer on membrane lipid rafts induces PI3K-Akt signaling pathway activation and then internalization [46]. C2 toxin is trafficked to EEs, where the release of C2I into the cytosol occurs. C2I translocation is promoted by the actions of Hsp90 and PPIase, as well as translocating Ia [49,50]. In the cytoplasm, C2I catalyzes the ADP-ribosylation of G-actin, ensuring the depolymerization of filamentous actin and the rounding up of sensitive cells. C2IIa is transferred to LEs and lysosomes [51]. Conversely, a portion of C2IIa is transported to REs. C2IIa reappearing on the cell membrane could re-associate with C2I. These findings demonstrate that C2IIa is endocytosed and sorted from EEs to REs or LEs and lysosomes [51].

4. Conclusions

Iota-toxin and C2 toxin belong to the bacterial AB toxin family. These toxins possess the potency to internalize into cells and to release an enzymatic component into the cytoplasmic space. The amino acid sequence identity of the binding component of iota toxin and C2 toxin is rather low. This difference is reflected in differences in receptors. Iota toxin receptor is a proteinaceous receptor such as LSR receptor and CD44. C2 toxin is a sugar receptor. The invasion process of both toxins has much in common. Pores formed by oligomers of binding components promote the release of enzymatic components from EEs into the cytoplasmic space. Iota-toxin and C2 toxin may become important tools to induce the entry of efficacious substances or targeted therapeutic agents into particular cells. On the other hand, inhibitors of internalization and intracellular trafficking have the potential for use as useful therapeutic treatments for infectious diseases.

Acknowledgments: This work was supported by JSPS KAKENHI Grant Number JP16K08794.

Author Contributions: M.T., T.T., S.S., M.O., Y.S., J.H., S.O., K.K. and M.N. conceived and designed the review article. M.N. wrote the paper.

Conflicts of Interest: The authors declare no conflict of interest.

References

1. Gibert, M.; Petit, L.; Raffestin, S.; Okabe, A.; Popoff, M.R. *Clostridium perfringens* iota-toxin requires activation of both binding and enzymatic components for cytopathic activity. *Infect. Immun.* **2000**, *68*, 3848–3853. [CrossRef] [PubMed]

2.   Popoff, M.R.; Boquet, P. Clostridial toxins. *Future Microbiol.* **2009**, *4*, 1021–1064. [CrossRef] [PubMed]
3.   Sakurai, J.; Nagahama, M.; Oda, M.; Tsuge, H.; Kobayashi, K. *Clostridium perfringens* iota-toxin: Structure and function. *Toxins* **2009**, *1*, 208–228. [CrossRef] [PubMed]
4.   Stiles, B.G.; Wilkins, T.D. Purification and characterization of *Clostridium perfringens* iota-oxin: Dependence on two nonlinked proteins for biological activity. *Infect. Immun.* **1986**, *54*, 683–688. [PubMed]
5.   Gülke, I.; Pfeifer, G.; Liese, J.; Fritz, M.; Hofmann, F.; Aktories, K.; Barth, H. Characterization of the enzymatic component of the ADP-ribosyltransferase toxin. CDTa from *Clostridium difficile*. *Infect. Immun.* **2001**, *69*, 6004–6011. [CrossRef] [PubMed]
6.   Popoff, M.R.; Boquet, P. *Clostridium spiroforme* toxin is a binary toxin which ADP-ribosylates cellular actin. *Biochem. Biophys. Res. Commun.* **1988**, *152*, 1361–1368. [CrossRef]
7.   Aktories, K.; Bärmann, M.; Ohishi, I.; Tsuyama, S.; Jakobs, K.H.; Habermann, E. Botulinum C2 toxin ADP-ribosylates actin. *Nature* **1986**, *322*, 390–392. [CrossRef] [PubMed]
8.   Nagahama, M.; Nagayasu, K.; Kobayashi, K.; Sakurai, J. Binding component of *Clostridium perfringens* iota-toxin induces endocytosis in Vero cells. *Infect. Immun.* **2002**, *70*, 1909–1914. [CrossRef] [PubMed]
9.   Nagahama, M.; Yamaguchi, A.; Hagiyama, T.; Ohkubo, N.; Kobayashi, K.; Sakurai, J. Binding and internalization of *Clostridium perfringens* iota-toxin in lipid rafts. *Infect. Immun.* **2004**, *72*, 3267–3275. [CrossRef] [PubMed]
10. Aktories, K.; Lang, A.E.; Schwan, C.; Mannherz, H.G. Actin as target for modification by bacterial protein toxins. *FEBS J.* **2011**, *278*, 4526–4543. [CrossRef] [PubMed]
11. Knapp, O.; Benz, R.; Popoff, M.R. Pore-forming activity of clostridial binary toxins. *Biochim. Biophys. Acta* **2016**, *1858*, 512–525. [CrossRef] [PubMed]
12. Mauss, S.; Chaponnier, C.; Just, I.; Aktories, K.; Gabbiani, G. ADP-ribosylation of actin isoforms by *Clostridium botulinum* C2 toxin and *Clostridium perfringens* iota toxin. *Eur. J. Biochem.* **1990**, *194*, 237–241. [CrossRef] [PubMed]
13. Gibert, M.; Monier, M.N.; Ruez, R.; Hale, M.L.; Stiles, B.G.; Benmerah, A.; Johannes, L.; Lamaze, C.; Popoff, M.R. Endocytosis and toxicity of clostridial binary toxins depend on a clathrin-independent pathway regulated by Rho-GDI. *Cell. Microbiol.* **2011**, *13*, 154–170. [CrossRef] [PubMed]
14. Songer, J.G. Clostridial enteric diseases of domestic animals. *Clin. Microbiol. Rev.* **1996**, *9*, 216–234. [PubMed]
15. Li, J.; Adams, V.; Bannam, T.L.; Miyamoto, K.; Garcia, J.P.; Uzal, F.A.; Rood, J.I.; McClane, B.A. Toxin plasmids of *Clostridium perfringens*. *Microbiol. Mol. Biol. Rev.* **2013**, *77*, 208–333. [CrossRef] [PubMed]
16. Tsuge, H.; Nagahama, M.; Nishimura, H.; Hisatsune, J.; Sakaguchi, Y.; Itogawa, Y.; Katunuma, N.; Sakurai, J. Crystal structure and site-directed mutagenesis of enzymatic components from *Clostridium perfringens* iota-toxin. *J. Mol. Biol.* **2003**, *325*, 471–483. [CrossRef]
17. Tsuge, H.; Nagahama, M.; Oda, M.; Iwamoto, S.; Utsunomiya, H.; Marquez, V.E.; Katunuma, N.; Nishizawa, M.; Sakurai, J. Structural basis of actin recognition and arginineADP-ribosylation by *Clostridium perfringens* iota-toxin. *Proc. Natl. Acad. Sci. USA* **2008**, *105*, 7399–7404. [CrossRef] [PubMed]
18. Tsurumura, T.; Tsumori, Y.; Qiu, H.; Oda, M.; Sakurai, J.; Nagahama, M.; Tsuge, H. Arginine ADP-ribosylation mechanism based on structural snapshots of iota-toxin and actin complex. *Proc. Natl. Acad. Sci. USA* **2013**, *110*, 4267–4272. [CrossRef] [PubMed]
19. Barth, H.; Aktories, K.; Popoff, M.R.; Stiles, B.G. Binary bacterial toxins: Biochemistry, biology, and applications of common *Clostridium* and *Bacillus* proteins. *Microbiol. Mol. Biol. Rev.* **2004**, *68*, 373–402. [CrossRef] [PubMed]
20. Papatheodorou, P.; Carette, J.E.; Bell, G.W.; Schwan, C.; Guttenberg, G.; Brummelkamp, T.R.; Aktories, K. Lipolysis-stimulated lipoprotein receptor (LSR) is the host receptor for the binary toxin *Clostridium difficile* transferase (CDT). *Proc. Natl. Acad. Sci. USA* **2011**, *108*, 16422–16427. [CrossRef] [PubMed]
21. Schmidt, G.; Papatheodorou, P.; Aktories, K. Novel receptors for bacterial protein toxins. *Curr. Opin. Microbiol.* **2015**, *23*, 55–61. [CrossRef] [PubMed]
22. Krug, S.M.; Hayaishi, T.; Iguchi, D.; Watari, A.; Takahashi, A.; Fromm, M.; Nagahama, M.; Takeda, H.; Okada, Y.; Sawasaki, T.; et al. Angubindin-1, a novel paracellular absorption enhancer acting at the tricellular tight junction. *J. Control. Release* **2017**, *260*, 1–11. [CrossRef] [PubMed]
23. Wigelsworth, D.J.; Ruthel, G.; Schnell, L.; Herrlich, P.; Blonder, J.; Veenstra, T.D.; Carman, R.J.; Wilkins, T.D.; Van Nhieu, G.T.; Pauillac, S.; et al. CD44 promotes intoxication by the clostridial iota-family toxins. *PLoS ONE* **2012**, *7*, e51356. [CrossRef] [PubMed]

24. Mañes, S.; del Real, G.; Martínez-A, C. Pathogens: Raft hijackers. *Nat. Rev. Immunol.* **2003**, *3*, 557–568. [CrossRef] [PubMed]
25. Zaas, D.W.; Duncan, M.; Wright, J.R.; Abraham, S.N. The role of lipid rafts in the pathogenesis of bacterial infections. *Biochim. Biophys. Acta* **2005**, *1746*, 305–313. [CrossRef] [PubMed]
26. Riethmüller, J.; Riehle, A.; Grassmé, H.; Gulbins, E. Membrane rafts in host-pathogen interactions. *Biochim. Biophys. Acta* **2006**, *1758*, 2139–2147. [CrossRef] [PubMed]
27. Hale, M.L.; Marvaud, J.C.; Popoff, M.R.; Stiles, B.G. Detergent-resistant membrane microdomains facilitate Ib oligomer formation and biological activity of *Clostridium perfringens* iota-toxin. *Infect. Immun.* **2004**, *72*, 2186–2193. [CrossRef] [PubMed]
28. Marvaud, J.C.; Smith, T.; Hale, M.L.; Popoff, M.R.; Smith, L.A.; Stiles, B.G. *Clostridium perfringens* iota-toxin: Mapping of receptor binding and Ia docking domains on Ib. *Infect. Immun.* **2001**, *69*, 2435–2441. [CrossRef] [PubMed]
29. Blonder, J.; Hale, M.L.; Chan, K.C.; Yu, L.R.; Lucas, D.A.; Conrads, T.P.; Zhou, M.; Popoff, M.R.; Issaq, H.J.; Stiles, B.G.; et al. Quantitative profiling of the detergent-resistant membrane proteome of iota-b toxin induced vero cells. *J. Proteome Res.* **2005**, *4*, 523–531. [CrossRef] [PubMed]
30. Papatheodorou, P.; Hornuss, D.; Nölke, T.; Hemmasi, S.; Castonguay, J.; Picchianti, M.; Aktories, K. *Clostridium difficile* binary toxin CDT induces clustering of the lipolysis-stimulated lipoprotein receptor into lipid rafts. *mBio* **2013**, *4*, e00244-13. [CrossRef] [PubMed]
31. Gibert, M.; Marvaud, J.C.; Pereira, Y.; Hale, M.L.; Stiles, B.G.; Boquet, P.; Lamaze, C.; Popoff, M.R. Differential requirement for the translocation of clostridial binary toxins: Iota toxin requires a membrane potential gradient. *FEBS Lett.* **2007**, *581*, 1287–1296. [CrossRef] [PubMed]
32. Nagahama, M.; Umezaki, M.; Tashiro, R.; Oda, M.; Kobayashi, K.; Shibutani, M.; Takagishi, T.; Ishidoh, K.; Fukuda, M.; Sakurai, J. Intracellular trafficking of *Clostridium perfringens* iota-toxin b. *Infect. Immun.* **2012**, *80*, 3410–3416. [CrossRef] [PubMed]
33. Kaiser, E.; Kroll, C.; Ernst, K.; Schwan, C.; Popoff, M.; Fischer, G.; Buchner, J.; Aktories, K.; Barth, H. Membrane translocation of binary actin-ADP-ribosylating toxins from *Clostridium difficile* and *Clostridium perfringens* is facilitated by cyclophilin A and Hsp90. *Infect. Immun.* **2011**, *79*, 3913–3921. [CrossRef] [PubMed]
34. Ernst, K.; Liebscher, M.; Mathea, S.; Granzhan, A.; Schmid, J.; Popoff, M.R.; Ihmels, H.; Barth, H.; Schiene-Fischer, C. A novel Hsp70 inhibitor prevents cell intoxication with the actin ADP-ribosylating *Clostridium perfringens* iota toxin. *Sci. Rep.* **2016**, *6*, 20301. [CrossRef] [PubMed]
35. Ernst, K.; Schmid, J.; Beck, M.; Hägele, M.; Hohwieler, M.; Hauff, P.; Ückert, A.K.; Anastasia, A.; Fauler, M.; Jank, T.; et al. Hsp70 facilitates trans-membrane transport of bacterial ADP-ribosylating toxins into the cytosol of mammalian cells. *Sci. Rep.* **2017**, *7*, 2724. [CrossRef] [PubMed]
36. Knapp, O.; Benz, R.; Gibert, M.; Marvaud, J.C.; Popoff, M.R. Interaction of *Clostridium perfringens* iota-toxin with lipid bilayer membranes. Demonstration of channel formation by the activated binding component Ib and channel block by the enzyme component Ia. *J. Biol. Chem.* **2002**, *277*, 6143–6152. [CrossRef] [PubMed]
37. Richard, J.F.; Mainguy, G.; Gibert, M.; Marvaud, J.C.; Stiles, B.G.; Popoff, M.R. Transcytosis of iota-toxin across polarized CaCo-2 cells. *Mol. Microbiol.* **2002**, *43*, 907–917. [CrossRef] [PubMed]
38. Nagahama, M.; Umezaki, M.; Oda, M.; Kobayashi, K.; Tone, S.; Suda, T.; Ishidoh, K.; Sakurai, J. *Clostridium perfringens* iota-toxin b induces rapid cell necrosis. *Infect. Immun.* **2011**, *79*, 4353–4360. [CrossRef] [PubMed]
39. Schleberger, C.; Hochmann, H.; Barth, H.; Aktories, K.; Schulz, G.E. Structure and action of the binary C2 toxin from *Clostridium botulinum*. *J. Mol. Biol.* **2006**, *364*, 705–715. [CrossRef] [PubMed]
40. Eckhardt, M.; Barth, H.; Blöcker, D.; Aktories, K. Binding of *Clostridium botulinum* C2 toxin to asparagine-linked complex and hybrid carbohydrates. *J. Biol. Chem.* **2000**, *275*, 2328–2334. [CrossRef] [PubMed]
41. Nagahama, M.; Takehara, M.; Takagishi, T.; Seike, S.; Miyamoto, K.; Kobayashi, K. Cellular uptake of *Clostridium botulinum* C2 toxin requires acid sphingomyelinase activity. *Infect. Immun.* **2017**, *85*, e00966-16. [CrossRef] [PubMed]
42. Idone, V.; Tam, C.; Andrews, N.W. Two-way traffic on the road to plasma membrane repair. *Trends Cell Biol.* **2008**, *18*, 552–559. [CrossRef] [PubMed]
43. Los, F.C.; Randis, T.M.; Aroian, R.V.; Ratner, A.J. Role of pore-forming toxins in bacterial infectious diseases. *Microbiol. Mol. Biol. Rev.* **2013**, *77*, 173–207. [CrossRef] [PubMed]

44. Tam, C.; Idone, V.; Devlin, C.; Fernandes, M.C.; Flannery, A.; He, X.; Schuchman, E.; Tabas, I.; Andrews, N.W. Exocytosis of acid sphingomyelinase by wounded cells promotes endocytosis and plasma membrane repair. *J. Cell Biol.* **2010**, *189*, 1027–1038. [CrossRef] [PubMed]

45. Corrotte, M.; Fernandes, M.C.; Tam, C.; Andrews, N.W. Toxin pores endocytosed during plasma membrane repair traffic into the lumen of MVBs for degradation. *Traffic* **2012**, *13*, 483–494. [CrossRef] [PubMed]

46. Nagahama, M.; Hagiyama, T.; Kojima, T.; Aoyanagi, K.; Takahashi, C.; Oda, M.; Sakaguchi, Y.; Oguma, K.; Sakurai, J. Binding and internalization of *Clostridium botulinum* C2 toxin. *Infect. Immun.* **2009**, *77*, 5139–5148. [CrossRef] [PubMed]

47. Gómez-Muñoz, A. Ceramide 1-phosphate/ceramide, a switch between life and death. *Biochim. Biophys. Acta* **2006**, *1758*, 2049–2056. [CrossRef] [PubMed]

48. Rivera, I.G.; Ordoñez, M.; Presa, N.; Gomez-Larrauri, A.; Simón, J.; Trueba, M.; Gomez-Muñoz, A. Sphingomyelinase D/ceramide 1-phosphate in cell survival and inflammation. *Toxins* **2015**, *7*, 1457–1466. [CrossRef] [PubMed]

49. Barth, H. Exploring the role of host cell chaperones/PPIases during cellular up-take of bacterial ADP-ribosylating toxins as basis for novel pharmacological strategies to protect mammalian cells against these virulence factors. *Naunyn-Schmiedebergs Arch. Pharmacol.* **2011**, *383*, 237–245. [CrossRef] [PubMed]

50. Kaiser, E.; Böhm, N.; Ernst, K.; Langer, S.; Schwan, C.; Aktories, K.; Popoff, M.; Fischer, G.; Barth, H. FK506-binding protein 51 interacts with *Clostridium botulinum* C2 toxin and FK506 inhibits membrane translocation of the toxin in mammalian cells. *Cell. Microbiol.* **2012**, *14*, 1193–1205. [CrossRef] [PubMed]

51. Nagahama, M.; Takahashi, C.; Aoyanagi, K.; Tashiro, R.; Kobayashi, K.; Sakaguchi, Y.; Ishidoh, K.; Sakurai, J. Intracellular trafficking of *Clostridium botulinum* C2 toxin. *Toxicon* **2014**, *82*, 76–82. [CrossRef] [PubMed]

toxins

MDPI

Review

Structural Features of Apicomplexan Pore-Forming Proteins and Their Roles in Parasite Cell Traversal and Egress

Alfredo J. Guerra and Vern B. Carruthers *

Department of Microbiology and Immunology, University of Michigan, Ann Arbor, MI 48109-5620, USA;
galfredo@umich.edu
* Correspondence: vcarruth@umich.edu; Tel.: +1-734-763-2081

Academic Editor: Alexey S. Ladokhin
Received: 2 August 2017; Accepted: 22 August 2017; Published: 29 August 2017

Abstract: Apicomplexan parasites cause diseases, including malaria and toxoplasmosis, in a range of hosts, including humans. These intracellular parasites utilize pore-forming proteins that disrupt host cell membranes to either traverse host cells while migrating through tissues or egress from the parasite-containing vacuole after replication. This review highlights recent insight gained from the newly available three-dimensional structures of several known or putative apicomplexan pore-forming proteins that contribute to cell traversal or egress. These new structural advances suggest that parasite pore-forming proteins use distinct mechanisms to disrupt host cell membranes at multiple steps in parasite life cycles. How proteolytic processing, secretion, environment, and the accessibility of lipid receptors regulate the membranolytic activities of such proteins is also discussed.

Keywords: apicomplexan; parasite; pore-forming proteins; membrane disruption; cell traversal; egress; protein structure; regulation

1. Introduction

Apicomplexans are obligate intracellular protozoan parasites that share a common set of apical secretory and cytoskeletal structures known as the apical complex. Notable members of the phylum Apicomplexa include *Cryptosporidium* spp., *Plasmodium* spp., and *Toxoplasma gondii*, which cause human cryptosporidiosis, malaria, and toxoplasmosis, respectively [1–3]. Other apicomplexans, such as *Eimeria* spp. and *Neospora caninum*, are important disease-causing agents in livestock, including poultry and cattle [4,5]. Apicomplexans replicate within a membrane-bound compartment, the parasitophorous vacuole (PV), in parasitized cells. Parasite escape from the PV after replication results in the cytolytic death of the host cell, causing the destruction of infected tissues and inducing inflammation, which together are the basis of the symptoms and disease. These parasites collectively exert a sizable toll on human and animal health, along with having a considerable economic effect.

Plasmodium and *Toxoplasma* are the most intensively studied apicomplexans because of their impact on human health, culturability, genetic tractability, and the availability of animal models for infection and disease [6]. These parasites have complex life cycles, involving an intermediate host for asexual replication and a definitive host wherein sexual replication culminates in a transmissible form. The infection of intermediate and definitive hosts requires substantial versatility by the parasite during its development in different types of host cells, along with an obligatory migration to distinct tissues to ensure efficient dissemination and transmission. As such, these parasites have evolved effective means for disrupting biological barriers, including membranes that confine or obstruct the parasites from progressing through their hosts.

Pore-Forming Proteins (PFPs) are often considered toxins because of their ability to create lesions in biological membranes, resulting in damage that compromises cell and tissue viability [7]. Many PFPs are synthesized in an inactive form and are secreted initially as soluble proteins that subsequently bind to target membranes and reconfigure or assemble into integral membrane pores [8]. Some PFPs function as portals to allow the translocation of other proteins and molecules, but the actions of many such proteins leads to the direct or indirect destruction of biological membranes [9,10]. PFPs can be binned into α- or β-PFPs, depending on whether membrane integration involves the insertion of α-helices or β-sheets.

During apicomplexan infection, PFPs have been implicated mainly in two events termed cell traversal and egress (Figure 1). Both events involve the concerted action of PFPs and parasite gliding motility driven by a linear actomyosin motor system lying directly beneath the parasite plasma membrane. Cell traversal, which is crucial for migrating through tissues, involves the gliding-dependent penetration of the parasite into a target host cell [11,12]. During penetration, the parasite either forms a transient vacuole or it traverses the host plasma membrane (HPM) from the outside-in, rendering it free in the cytosol (Figure 1, top). Parasites residing in a transient vacuole exit in two steps: disruption of the transient vacuole (akin to the outside-in step) and inside-out traversal of the HPM [13]. A cytosolic parasite exits in a single step by directly rupturing the HPM from the inside-out. Egress, which is necessary to liberate the parasite after replication or development, occurs via the gliding- and PFP-dependent disruption of the PV and HPM (Figure 1, bottom) [14,15]. Cell traversal and egress are similar in that both involve parasite disruption of host-derived membrane barriers, but are distinguished by the timing of exit, which occurs within seconds for cell traversal and hours or days for egress.

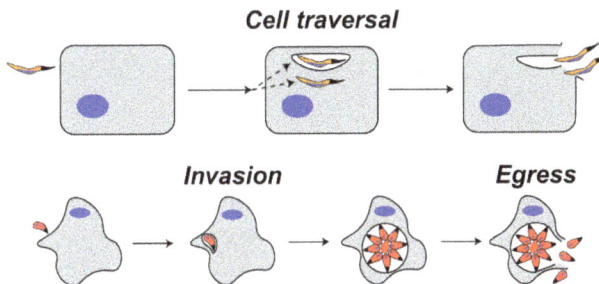

Figure 1. Representation of cell traversal (top) versus egress (bottom) by apicomplexans. During cell traversal, a sporozoite enters the host cell and is either contained in a transient vacuole or remains free in the cytosol. The parasite then exits the host cell by rupturing the transient vacuole and the host plasma membrane (HPM) or by traversing the HPM from the cytosol. During invasion, the parasite enters the cell and forms a parasitophorous vacuole (PV) where replication occurs. After replication, parasites egress from the PV and the host cell by rupturing the PV membrane and the HPM.

Although the molecular and cellular underpinnings of cell traversal and egress remain poorly understood, recent crystallographic studies of known or putative apicomplexan PFPs have yielded important new insight into how such proteins contribute to cell traversal or egress. This review highlights the structural and functional advances for *Plasmodium* α-PFPs and discusses the role of membrane attack complex/perforin (MACPF) β-PFPs in *Plasmodium* and *Toxoplasma* infection biology.

2. Malaria Encodes Two PFPs That Are Unique among Apicomplexans

To provide context for the stages relevant to PFP function, we briefly describe the life cycles of *Plasmodium* and *Toxoplasma* in this and subsequent sections. Further discussion of the life cycle can be found elsewhere [16,17]. The *Plasmodium* infection of a mammalian host begins when a sporozoite is injected into the host by an infected mosquito (Figure 2). Sporozoites are highly motile and migrate through the dermis to enter the blood stream and migrate to the liver. Once in the liver, sporozoites breach the liver sinusoid layer by traversing through Kupffer cells and hepatocytes until they create a productive PV. Sporozoites traverse cells by disrupting membranes to move out of the cell, a process first described in Kupffer cells with the mouse malaria parasite *Plasmodium berghei* [18]. Similar to other *Plasmodium* spp., *P. berghei* encodes three proteins that are required for cell traversal: (1) Sporozoite microneme Protein Essential for Cell Traversal 1 (SPECT1); (2) Sporozoite microneme Protein Essential for Cell Traversal 2 (SPECT2; also known as Perforin-Like Protein 1 (PLP1)); and (3) Cell-Traversal protein for Ookinetes and Sporozoites (CelTOS) (Table 1). In addition to functioning in sporozoite cell traversal, CelTOS, as the name implies, is also required for ookinete cell traversal in the mosquito stage. *Plasmodium* parasites additionally encode four other Perforin-Like Proteins (PLP2-5), some of which also play a role in ookinete cell traversal. All of these proteins are released from the parasite micronemes, which are calcium-responsive apical secretory organelles that are discharged during cell traversal, egress, and invasion [19]. In the following sections, we will discuss SPECT1 and CelTOS, since these two proteins are unique to *Plasmodium* parasites. PLP proteins will be discussed more extensively in later sections.

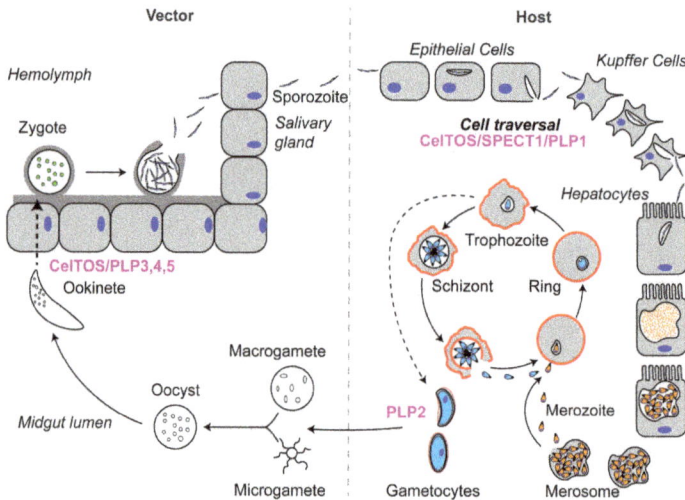

Figure 2. Schematic representation of the *Plasmodium* spp. life cycle. See text for descriptions of Pore-Forming Proteins (PFPs) and their roles in each step of the life cycle. PFPs are indicated in magenta.

Table 1. Proteins involved in apicomplexan cell traversal or egress.

Protein	Parasites	Type of PFP	*Toxoplasma* Stage/Function	*Plasmodium* Stage/Function
SPECT1	*Plasmodium* spp.	unknown	NA	sporozoite/cell traversal
CelTOS	Aconoidasida	α [4]	NA	sporozoite and ookinete/cell traversal
PLP1/SPECT2	Apicomplexa [1]	β	tachyzoite/egress	sporozoite/cell traversal, merozoite/egress
PLP2	Apicomplexa [1]	β	sexual stages/unknown	male gametocyte/egress, merozoite/egress
PLP3//MAOP	Apicomplexa [2]	β	NA	ookinete/cell traversal
PLP4	Apicomplexa [3]	β	NA	ookinete/cell traversal
PLP5	Apicomplexa [3]	β	NA	ookinete/cell traversal

[1] Not present in *Cryptosporidium* or *Gregarina*. [2] Not present in *Cryptosporidium*, *Gregarina*, *Toxoplasma*, *Cyclospora*, or *Eimeria*. [3] Only present in Aconoidasida and *Cytauxzoon*. [4] Implied based on the crystal structure.

2.1. SPECT1: A Unique Plasmodium Protein for Cell Traversal

2.1.1. SPECT1 and Cell Traversal

P. berghei sporozoites lacking SPECT1 (*Pb∆spect1*) have wild-type-like gliding motility and hepatocyte infectivity, but are unable to traverse sinusoidal layer cells and thus incapable of accessing hepatocytes in the parenchyma of the liver [20]. Treatment with liposomal clondronate, which depletes Kupffer cells, restored normal levels of infection presumably due to the ability of the parasites to glide through the gaps formed by Kupffer cell depletion. *Pb∆spect1* parasites are 5- to 10-fold less infective than wild-type parasites when injected subcutaneously. Additionally, most *Pb∆spect1* sporozoites are trapped in mouse dermis after deposition from a mosquito bite, despite having normal gliding motility in three-dimensional (3D) matrices [12]. Recent work has also shown that *Plasmodium falciparum* SPECT1 (*Pf*SPECT1) contributes to *P. falciparum* infection of mice with humanized liver tissue [21]. SPECT1 is highly conserved among *Plasmodium* species (~40% sequence identity), consistent with sharing a common structure and function. Indeed, BLAST searches querying apicomplexan genomes with SPECT1 only return matches within the *Plasmodium* genus.

2.1.2. P. berghei SPECT1 Crystal Structure: A Common Structural Fold Associated with Membrane Proteins

Recent solving of the 2.7 Å crystal structure of *Pb*SPECT1 provided some intriguing, albeit cryptic, potential clues to its function in cell traversal [22]. The structure shows a nearly parallel four α-helix bundle with a "hook" feature on one end (Figure 3). The four-helix bundle is a fairly common structural fold found in a subset of transmembrane proteins. Accordingly, a Dali Lite server [23] search with the *Pb*SPECT1 structure yields similarity to a THATCH domain of Huntingtin-interacting protein 1 (HIP1R THATCH, PDB code 1R0D) [24], a coiled-coil domain of a Potato Virus X Resistance Protein (RxCC, PDB code 4m70) [25], a transmembrane domain of a eukaryotic V-type ATPase (PDB code 5VOZ) [26], and a yeast t-SNARE protein (Sso2, PDB code 5M4Y) [27]. The last two of these proteins are either transmembrane domains or have membrane-binding activity. However, no transmembrane helices are predicted in the *Pb*SPECT1 sequence, and it remains unclear if or how SPECT1 interacts with membranes during *Plasmodium* cell traversal. One enticing feature of the crystal structure is a rather deep pocket flanked by the α-3 and α-4 helices (Figure 3c). This cavity is lined primarily with hydrophobic residues that are conserved in other SPECT1 homologs, and could be a binding site for cholesterol or lipids. However, another plausible interpretation of this cavity and the parallel α-helices is that they indicate a measure of instability in the crystallized structure. This could be telling of potential conformational changes that may occur in SPECT1. Interestingly, the structure of *Pb*SPECT1 also bears a striking similarity to the N-terminal portion of a mammalian protein termed Izumo1, which also consists of a nearly parallel four α-helix bundle with a hook feature [28,29]. This so-called Izumo domain is crucial to the function of Izumo1 in the fusion of mammalian sperm with an egg. SPECT1 and the Izumo domain share solvent-exposed bulk hydrophobic amino acids speculated to interface with other proteins. Although additional studies are necessary to understand how SPECT1 and the Izumo domain confer membrane traversal or fusion, these crystallographic studies provide an important basis for future work identifying associated proteins and mechanisms underlying membrane interaction.

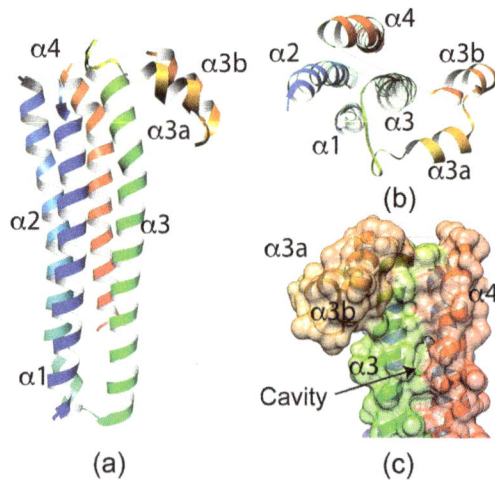

Figure 3. (**a**) Side view of the crystal structure of *Plasmodium berghei* SPECT1. The structure shows a four-helix bundle with a hook-like element between α3 and α4. (**b**) Top view of the SPECT1 crystal structure. The dashed lines between α1–α2 and α3b–α4 represent loops lacking electron density. (**c**) Surface representation of the back of the SPECT1 crystal structure showing the cavity formed between α3 and α4.

2.2. CelTOS: A Conserved Protein in the Aconoidasida Is Required for Cell Traversal in Both the Mammalian Host and Vector

2.2.1. CelTOS Dictates Cell Traversal by Sporozoites and Ookinetes

CelTOS is unique to the Aconoidasida class of apicomplexan parasites, which includes *Plasmodium* spp. and the tick borne parasites *Babesia* spp. and *Theleria* spp. Genetic studies in *Plasmodium* established the essentiality of CelTOS for efficient traversal in both the mammalian host (sporozoite) and mosquito vector (ookinete). Indeed, CelTOS is only expressed by ookinetes and salivary gland sporozoites, but not by immature sporozoites in the oocyst [30]. Whereas Δ*celtos* parasites show normal ookinete development in vivo or in vitro, they exhibit a marked defect in migration through midgut epithelial cells, resulting in 200-fold reduction in subsequent oocysts. Similarly, Δ*celtos* sporozoites show normal microneme secretion, but have reduced infectivity based on exhibiting a longer prepatent period upon the infection of mice. This phenotype is reversed when sporozoites lacking CelTOS are inoculated in Kupffer cell-depleted rats. Finally, despite having a similar ability to form PVs in a HEPG2 cell line, Δ*celtos* sporozoites have much lower cell wounding activity. Taken together, these experiments highlight the importance of CelTOS in both ookinete and sporozoite cell traversal, rendering this protein a potential vaccine target for preventing liver infection [31,32].

2.2.2. CelTOS Is a PFP with a Preference for Phosphatidic Acid

Despite its importance as a cell traversal protein and being a conserved protein within a subset of apicomplexan parasites, the mechanism by which CelTOS aides in cell traversal has remained elusive largely due to a lack of similarity to other proteins of known function. Recent solving of the *Plasmodium vivax* CelTOS (*Pv*CelTOS) 3.0 Å crystal structure revealed a highly α-helical dimer resembling a tuning fork (Figure 4) [33]. Although the overall structure lacks similarity to canonical pore-forming proteins, the N-terminal and C-terminal subunits bear some resemblance to the membrane disruptive viral and bacterial proteins HIV-gp41, Hendravirus fusion protein, Nipahvirus fusion protein, and *Mycobacterium bovis* ESAT-6 [33]. Additionally, both *Pf*CelTOS and *Pv*CelTOS bind phosphatidic acid (PA) in a spotted lipid array, and at nanomolar protein concentrations can disrupt liposomes containing PA [33]. Liposomes containing phosphatidylserine or phosphatidylcholine required micromolar concentrations of CelTOS. If PA is the physiologic receptor for CelTOS, the known localization of PA to the inner leaflet of the HPM implies a role for CelTOS in the inside-out disruption of the HPM for sporozoite exit. Finally, transmission electron microscopy of negative stained liposomes with and without CelTOS show the formation of CelTOS-dependent ~50 nm pores [33]. These studies provide important new insight into how CelTOS facilitates parasite cell traversal and lay a foundation for future work identifying the structural rearrangements associated with pore formation.

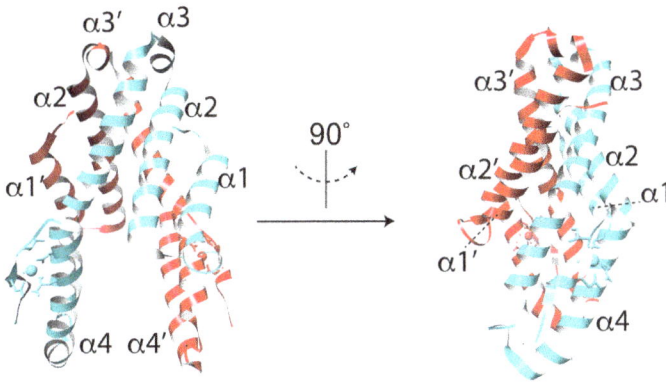

Figure 4. Crystal structure of the CelTOS dimer in ribbon representation with each monomer colored in cyan and red. The N-terminal subunit is defined by helices α1 and α2. The C-terminal subunit is defined by helices α3 and α4.

3. Apicomplexans Share a Family of MACPF PFPs that Function at Multiple Steps in Their Life Cycles

3.1. MACPF/CDC Proteins Have a Variable C-Terminal Domain

MACPF proteins are members of a large and diverse family of pore-forming proteins found in virtually all kingdoms of life, including apicomplexans [19,34,35]. The pore-forming MACPF domain, which has structural similarity to domains 1 and 3 of cholesterol-dependent cytolysins (CDCs), is defined by a central four-stranded antiparallel β-sheet (Figure 5a, cyan) and two clusters of α-helices termed CH1 and CH2 (Figure 5a, magenta). The first step in pore formation involves membrane recognition and binding. In CDCs, this occurs via the conserved domain 4 (Figure 5a, blue), which is comprised of an immunoglobulin-like fold that recognizes cholesterol and initiates binding to the membrane. MACPF proteins, on the other hand, have significant variation in the C-terminal domain (CTD), and the membrane receptors for these proteins remain unknown. The variability in CTD structures becomes evident when comparing the crystal structures for Perforin 1 (Figure 5b; C2 domain fold) and Plu-MACPF (Figure 5c; β-prism fold). Apicomplexan MACPF proteins, termed Perforin-like proteins (ApiPLPs), share the conserved MACPF domain and a downstream β-rich domain (Figure 5d). The β-rich domain is similar to the CTDs of CDC and other MACPF proteins in that it is predicted to consist mainly of β-pleated sheets. However, unlike other CTDs, the β-rich domain is comprised of three direct repeats, each containing four highly conserved cysteine residues that are presumed to form disulfide linkages. Several hydrophobic residues are also partially conserved. The extent to which the repeats form a single globular domain or three tandemly linked mini-domains remains unknown. The signature repeat is conserved among the β-rich domains of ApiPLPs, but otherwise the domain sequence has no homology to other proteins. Although an in silico molecular modeling study [36] suggested that the β-rich domain of *Plasmodium* PLPs resembles a C2 domain fold, C2 domains do not typically contain repeat elements and the C2 signature sequence (Pfam ID number PF00168) bears no resemblance to the β-rich motif. Thus, it remains to be determined whether the ApiPLP β-rich domain is structurally related to other CTDs of MACPF proteins such as the C2-domain of perforin. Regardless, after binding of a MACPF protein via its CTD to the target membrane, the protein oligomerizes into ring- or arc-shaped complexes. These so-called pre-pore complexes then undergo a dramatic structural rearrangement of the MACPF domain, wherein the CH1 and CH2 helices unfurl to become extensions of the central β-sheets, thereby knifing into the target membrane to form a pore as a super β-barrel. *T. gondii* PLP1 (*Tg*PLP1) and *P. falciparum* PLP1 (*Pf*PLP1) were shown to bind membranes, oligomerize, and have pore-forming activity [37–40]; thus, it appears that ApiPLPs share the basic mechanism of pore formation with other MACPF proteins.

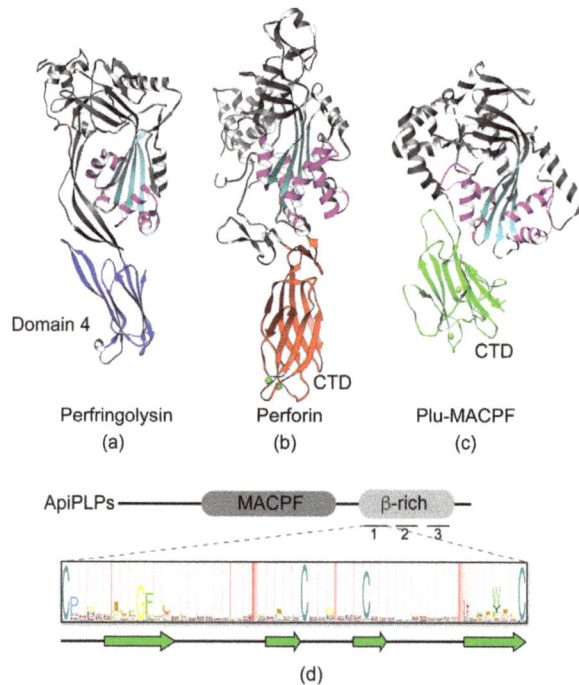

Figure 5. Representative structures of the membrane attack complex/perforin (MACPF)/ cholesterol-dependent cytolysins (CDC) family of proteins. The conserved MACPF/CDC domain is shown in cyan and two conserved helical clusters (CH1/CH2) are colored in magenta. (**a**) Structure of Perfingolysin (PDB code 1PFO) with domain 4 highlighted in blue. (**b**) Structure of Perforin 1 (PDB code 3NSJ) with the C-terminal domain (CTD) colored in red. Calcium ions in the CTD are shown as green spheres. (**c**) Structure of Plu-MACPF (PDB code 2QP2) with the C-terminal domain colored in green. Calcium ions in the CTD are shown as green spheres. (**d**) Diagram of the domain organization of a generic ApiPLP illustrating the MACPF and β-rich domains. The domain is made up of three direct repeats of ~60 amino acids, each containing conserved cysteine and hydrophobic amino acids illustrated in the Hidden Markov Model (HMM) logo created from 51 individual repeats representing 22 ApiPLPs (*Tg*PLP1-2, *Neospora caninum* NcPLP1-3, *Pf*PLP1-5, *Babesia bovis* BbPLP1-6; *Theileria annulata* TaPLP1-6). The degree of conservation at each position is proportional to the size of the single letter code indicating the residue. Approximate positions of predicted β-sheets are depicted below the HMM logo.

3.2. *Plasmodium MACPF Proteins Are Important for All Stages of the Parasite Life Cycle*

3.2.1. Role of *Plasmodium* PLP1 in Cell Traversal

As mentioned above, during the infection of the mammalian host *Plasmodium*, sporozoites must traverse a variety of cell types before developing into a non-motile, replicating exo-erythrocytic form (EEF) inside a hepatocyte (Figure 2). Cell traversal of dermal and liver sinusoidal layer cells is dependent on both SPECT1 and PLP1. Sequence alignments show the similarity of *Pf*PLP1 to other MACPF proteins, including a signature motif (Y/W-G-T/S-H-F/Y-X6-G-G) as well as helical clusters, CH1 and CH2. Previous work reported that *Pf*PLP1 was expressed in blood stages and mediates the calcium-dependent egress of merozoites from erythrocytes [39]. However, studies with *P. falciparum* parasites expressing a triple HA9 epitope endogenously tagged PLP1 failed to detect expression of *Pf*PLP1 during the blood stage [41]. Additionally, a deletion strain of *Pf*PLP1 (*Pf*Δ*plp1*) showed normal asexual growth when cultured in erythrocytes in vitro [41]. *Pf*Δ*plp1* parasites also are not defective in a number of gametocytes, midgut oocysts, or salivary gland sporozoites. *Pf*Δ*plp1* sporozoites are, however, deficient in the cell traversal of both hepatocytes and human monocyte-derived macrophages. This defect appears to be confined to cell traversal, since *Pf*Δ*plp1* parasites showed normal invasion of hepatocytes and EEF development. Finally, the intravenous inoculation of *Pf*Δ*plp1* sporozoites into mice with humanized liver tissue failed to establish infection, whereas wild-type parasites readily infect hepatocytes in this same system. These observations, which are consistent with earlier studies of *P. berghei* parasites lacking *Pb*PLP1 [12,42,43], suggest a principal role for *Plasmodium* PLP1 in cell traversal during sporozoite migration from the dermis to the blood stream and movement across the liver sinusoidal layer into parenchymal hepatocytes.

Initial work describing hepatocyte cell traversal by rodent *Plasmodium yoelii* sporozoites suggested a model in which the parasite crosses the HPM from the outside-in and it temporarily occupies the host cytosol before exiting the cell via an inside-out rupture of the HPM [11]. This model was based mainly on the observation by electron microscopy of sporozoites lacking a vacuolar membrane. Other work with *P. berghei* or *P. yoelii* sporozoites and Kupffer cells, however, observed sporozoites within a vacuole, putatively during the cell traversal of such cells [18,44]. Nevertheless, these hepatocyte and Kupffer cells studies did not directly observe the fate of the HPM or vacuole membrane during cell traversal. More recent work carefully examining temporal steps in cell traversal concluded that the majority of cell-traversing *P. yoelii* sporozoites occupy a transient vacuole in hepatocytes, and that escape from the transient vacuole requires the expression of *P. yoelii* PLP1 (*Py*PLP1) [13]. Failure to escape resulted in the fusion of hepatocyte lysosomes with the vacuole, leading to sporozoite death. These findings suggest an intriguing new facet of cell traversal involving a transient vacuole. Additional work is needed to define the extent to which transient vacuoles exist in vivo or are used by other *Plasmodium spp.*

3.2.2. *Plasmodium* PLP2 Functions in Gametocyte Egress

After exiting the liver stage, *Plasmodium* parasites infect erythrocytes and undergo a second round of asexual reproduction in the symptomatic erythrocytic stage (Figure 2). Generally, the erythrocytic cycle is separated into three developmental stages: ring, trophozoite, and schizont. However, a small percentage of trophozoites differentiate into gametocytes within erythrocytes, from which they must egress for mating in the mosquito midgut. Gametocyte egress is dependent on a second perforin-like protein (PLP2) expressed in male gametocytes [45]. A strain of *P. berghei* lacking PLP2 shows normal development of gametocytes in the blood stage, but the gametocytes are unable to egress and have aberrant exflagellation. This phenotype is overcome by treatment with saponin or the pore-forming toxin equinatoxin II prior to when erythrocytic membrane rupture would normally occur. Mutant *P. falciparum* parasites lacking PLP2 also showed normal egress during the intraerythrocytic cycle, but aberrant egress of gametocytes that is rescued by equinatoxin II treatment. These studies suggest that *Plasmodium* does not use MACPF proteins for cell traversal exclusively, but also deploys them for egress from infected cells in a manner akin to *T. gondii* (see below).

3.2.3. *Plasmodium* PLP3,4,5 and Cell Traversal in the Mosquito Midgut

Following the ingestion of a blood meal from an infected host, male and female *Plasmodium* gametocytes mate to generate a motile zygote known as an ookinete. Ookinetes must traverse the midgut epithelium to avoid mosquito immune defenses and access the basolateral membrane for further development into oocyts. This step is a major bottleneck in the lifecycle, which is likely why the parasite has evolved four individual PFPs to mediate the traversal of midgut epithelium. One of these, CelTOS, is described above. The other three PFPs are MACPF proteins termed PLP3 (also known as MAOP), PLP4, and PLP5.

P. berghei PLP3 (*Pb*PLP3) is expressed in the ookinete stage, where it localizes to micronemes. Parasites lacking *Pb*PLP3 (*Pb*Δ*plp3*) show a striking lack of infectivity in mosquitoes and are incapable of traversing epithelial cells. *Pb*PLP3-deficient ookinetes can attach to the apical surface of the midgut epithelium, but are unable to disrupt the HPM of epithelial cells [46].

P. falciparum ookinetes show a large increase in PLP4 (*Pf*PLP4) transcript levels compared to blood or gametocyte stages. *Pf*PLP4 deficient parasites have normal blood stage replication, gametocytogenesis, and exflagellation, suggesting that *Pf*PLP4 is not important during these life cycle events. However, feeding mosquitoes with the mature gametocytes of *Pf*Δ*plp4* parasites results in a significant reduction of oocysts when compared to a wild-type control, consistent with an inability to traverse midgut epithelial cells.

Similarly, a *P. berghei plp5* deletion strain (*Pb*Δ*plp5*) can bind to the apical side of the midgut epithelium, but is unable to traverse epithelial cells. However, this lack of infectivity is not absolute, since a small number of oocysts and salivary gland sporozoites were detected in infected mosquitoes. Interestingly, the direct injection of *Pb*Δ*plp5* ookinetes into the hemocoel, which effectively bypasses traversal of the midgut epithelium [47], restores the numbers of oocysts and salivary gland sporozoites, pinpointing the role of *Pb*PLP5 in the traversal of the midgut epithelium.

It is interesting to note that, despite the presence of four PFPs (CeLTOS, PLP3,4,5) during ookinete midgut epithelium invasion, there appears to be little functional overlap in this process. The possibility that pore formation by PLP3,4,5 might require heteromultimerization was proposed as a potential explanation for the apparent lack of functional redundancy [34]. These proteins might also act at different steps, namely the outside-in and inside-out traversal of the epithelial cell HPM, as suggested previously [48]. It should be noted that pore formation by PLP3,4,5 has not been established, but rather is inferred based on membership in the MACPF family. Further studies into the individual roles of CelTOS and PLP3, 4, and 5 are necessary to understand the mechanism by which malaria ookinetes traverse the midgut epithelium.

3.3. *Toxoplasma* MACPF Proteins

3.3.1. *Toxoplasma* Infection of Its Hosts

The *Toxoplasma gondii* life cycle is simpler than the *Plasmodium* life cycle, but still involves the infection of two different hosts. Again, we will only briefly discuss the life cycle to give context to our discussion of PFPs. The *Toxoplasma* life cycle can be generally divided into feline and non-feline infections (Figure 6). Sexual reproduction only occurs during the infection of felines, the definitive host. *T. gondii* is capable of infecting and replicating asexually within virtually all warm-blooded vertebrates, thus it is highly promiscuous for intermediate hosts. Infection begins with the consumption of a tissue cyst or a sporulated oocyst. During invasion, parasites form a PV wherein tachyzoites replicate asexually to form vacuoles containing 16 or more parasites. Tachyzoites then egress from the cell and invade neighboring host cells. This sequence of invasion, asexual reproduction, and egress constitutes the lytic cycle of acute infection. Following the acute infection, tachyzoites undergo differentiation into slow-growing bradyzoites, which develop into intracellular tissue cysts during chronic infection. If the tissue cyst is consumed by a non-feline intermediate host, the asexual reproductive cycle begins again. If a felid consumes the tissue cyst, the parasites undergo schizogony and sexual differentiation to form oocysts that are shed in the feces. In the following section, we will describe the role that *T. gondii* PLP1 (*Tg*PLP1) plays during egress and postulate a putative role for *T. gondii* PLP2 (*Tg*PLP2) during the sexual stage.

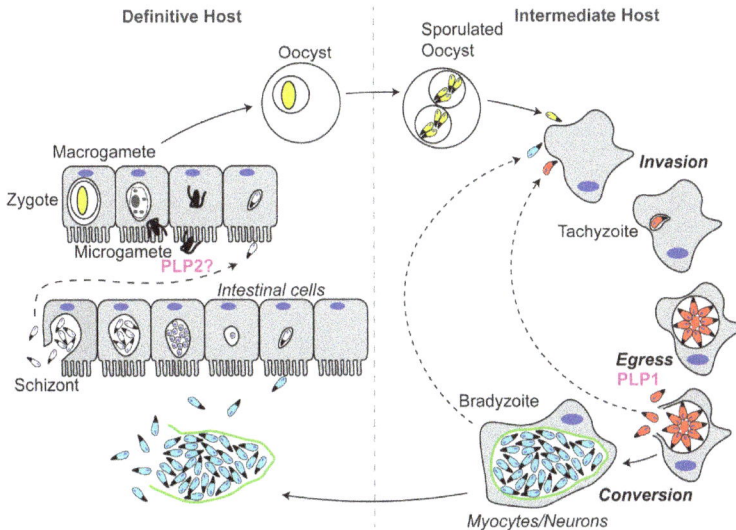

Figure 6. Schematic representation of the *Toxoplasma gondii* life cycle. PFPs that are important to the life cycle are labeled in magenta.

3.3.2. *Toxoplasma* PLP1 & PLP2

The *Toxoplasma gondii* genome encodes two PLP proteins, *Tg*PLP1 and *Tg*PLP2. A sequence analysis and homology modeling confirm that both proteins have the conserved features of a MACPF protein, including the signature sequence motif and the CH1 and CH2 regions in the pore-forming domain [15]. Interestingly, similar to the CTDs of MACPF family proteins, the C-terminal domains of *Tg*PLP1 and *Tg*PLP2 are predicted to be rich in β-pleated sheets, but they bear no sequence homology with other MACPF CTDs, as noted above. The *Tg*PLP1 CTD is required for pore-forming activity [37]. *Tg*PLP1 is secreted from micronemes in a calcium-dependent manner, and facilitates parasite egress by disrupting the PV membrane [15]. Video microscopy experiments showed that *Tg*Δ*plp1* parasites became motile within the PV after stimulating calcium signaling, but they show a substantial delay in crossing the PV membrane and exiting the host cell. *Tg*Δ*plp1* parasites also have a defect in spontaneous egress, based on observing large clusters of parasites trapped in "spherical structures" in host cells during routine culture. Further examination by electron microscopy revealed that these spheres were contained by both the vacuolar membrane as well as the host plasma membrane. Interestingly, mice infected with high doses of *Tg*PLP1-deficient parasites survive the infection, implicating *Tg*PLP1 as a key virulence factor [15]. In addition to the central MACPF domain and a β-rich CTD, *Tg*PLP1 has an N-terminal domain that has also been shown to have membrane-binding activity [37]. Although this domain is dispensable for pore formation and virulence, parasites expressing *Tg*PLP1 lacking the N-terminal domain egress less efficiently from host cells than those expressing the complete *Tg*PLP1, indicating a contributing but non-essential role in PLP1 function.

Transcriptional profiling (Toxodb.org) and epitope tagging [15] experiments suggest that *Tg*PLP2 is not expressed during acute asexual infection, and instead is produced during sexual development in the intestinal system of felids when *Tg*PLP1 is transcriptionally silent. As such, *Tg*PLP2 might play a role in gametocyte egress analogous to *Plasmodium* PLP2.

4. Regulation of Apicomplexan PFPs

Intracellular PFP activity is often limited in eukaryotic cells by producing the protein in a pro-form and by localization to a secretory organelle. For example, perforin 1 activity in cytotoxic T cells is controlled during intracellular transport by a C-terminal propeptide including a terminal tryptophan, which mediates rapid export from the ER, but is not required for cytolysis [49,50]. Deletion of the propeptide or mutation of Trp led to perforin 1 accumulation in the ER and cytotoxicity. *Tg*PLP1 and *Tg*PLP2 are likewise produced as pro-proteins with an N-terminal signal anchor sequence that is sufficient for *Tg*PLP1 targeting to micronemes [36]. Although the subcellular location at which the *Tg*PLP1 signal anchor propeptide is cleaved is not known, tethering *Tg*PLP1 to the luminal membrane by the N-terminal signal anchor may prevent autolysis during trafficking by preventing the C-terminal domain from binding the membrane. However, the mechanisms that suppress activity once maturation has occurred and the protein is stored in the micronemes remain to be determined. *Plasmodium* PLPs lack a signal anchor sequence and instead possess a signal sequence for translation into the ER lumen. It remains unclear if *Plasmodium* PLPs contain autoinhibitory sequences that suppress cytolytic activity during trafficking to and storage within the micronemes.

Storage within micronemes permits the regulated release of apicomplexan PFPs precisely when and where they are needed to disrupt biological membranes during cell traversal and egress. Calcium signaling within the parasite triggers the microneme release of PFPs along with activating parasite motility, which together deliver the cytolytic activity and mechanical force necessary for cell traversal and egress. How the parasite remains seemingly immune to the actions of its own PFPs is unknown, but prospectively could involve inhibitory proteins on the parasite surface or the lack of specific lipid or protein receptors.

Extracellular *Toxoplasma* tachyzoites secrete *Tg*PLP1 and other microneme proteins in response to a decrease in pH of the medium via activation of calcium signaling in the parasite [15,37]. In addition to stimulating the release of *Tg*PLP1, low pH also promotes the cytolytic activity of *Tg*PLP1 by enhancing membrane binding via its CTD [38]. The combined actions of increased release and enhanced activity in low pH medium rendered tachyzoites capable of wounding host cells in a *Tg*PLP1-dependent manner. Such activity is not normally seen in tachyzoites, which typically invade into a PV without damaging host cells. Additional work provided evidence on a population scale that the PV of intracellular dividing tachyzoites acidifies near the time of egress. Also, treatments that antagonize acidification impaired parasite egress from host cells upon stimulation of calcium signaling. Interestingly, the pretreatment of hepatocytes with bafilomycin A, which blocks V-type ATPase-dependent proton transport, trapped wild-type *Plasmodium* sporozoites in transient vacuoles [13]. Since escape from transient vacuoles is dependent upon the expression of *Plasmodium* PLP1, the findings imply that the acidification of the transient vacuole, possibly by the fusion of host lysosomes, promotes PLP1 secretion and/or activity. Together, these studies suggest a working model in which the acidification of *Toxoplasma*- and *Plasmodium*-containing vacuoles promotes PLP1-dependent membranolytic activity for egress or cell traversal, respectively. However, it should be noted that the pH of individual PVs or transient vacuoles prior to egress or escape has not been measured directly. Thus, additional work is needed to validate vacuole acidification and identify the underlying mechanism(s).

In addition to regulation by autoinhibitory sequences, storage in micronemes, and environmental pH, a fourth level of regulation is likely conferred by receptor availability, exemplified by CelTOS [33]. As mentioned above, PA is typically enriched in the inner leaflet of the HPM, thereby supporting a role for CelTOS in inside-out disruption of the HPM during cell traversal. In this scenario, CelTOS would not act optimally on an outside-in rupture of the HPM or transient vacuole, which instead is attributed to *Plasmodium* PLP1 activity. By extension, other apicomplexan PLPs may recognize leaflet-specific receptors that dictate the directionality of their actions. In this way, receptor accessibility could also contribute to the directionality of *Tg*PLP1's role in promoting tachyzoite egress from host cells without compromising the formation of an intact PV during subsequent invasion.

5. Perspectives

In this review, we have summarized recent insight gained from structural studies of apicomplexan PFPs and their roles in infection. Current evidence suggests that PFPs are important factors throughout the *Plasmodium* and *Toxoplasma* life cycles, but several important questions remain unanswered. First, despite the evidence that *Pf*SPECT1 is required for cell traversal, its activity as a PFP remains uncertain. Even with the crystal structure in hand, the mechanism by which *Pf*SPECT1 potentially forms pores remains unknown. In the same vein, the crystal structure of *Pf*CelTOS bears some resemblance to PFPs from other organisms, but it is unclear if or how *Pf*CelTOS recognizes PA during cell traversal. Furthermore, if PA is the cognate receptor for *Pf*CelTOS, there is a strong implication of protein activity in the latter portions of cell traversal, as PA is primarily an inner leaflet phospholipid; however, further work is necessary to confirm this hypothesis. Additionally, although there is evidence that *Pf*CelTOS forms pores, the extent of the structural rearrangements associated with pore formation remains obscure. High-resolution cryoelectron microscopy of *Pf*CelTOS has the potential to identify such structural rearrangements.

Despite our current knowledge of the structures of MACPF/CDC proteins in other organisms, we currently lack any empirical information on the structures of *P. falciparum* or *T. gondii* PLPs. This is particularly important considering the unique features of MACPF CTDs. The CTD is a key anchor point in the mechanism for pore formation. Solving the structures of the ApiPLPs will allow further understanding of how this set of proteins bind membranes prior to pore formation. Structural studies of ApiPLPs should also aid in understanding the basis for pH-dependent PLP activity. Further studies are also necessary to identify the receptors for ApiPLPs, since these may shed light on the directionality, i.e., inside-out or outside-in, of membrane disruption by these proteins. Finally, it is likely that there are other factors that influence parasite egress. Generally, the pores formed by the proteins described herein (50–200 Å or 5–20 nm) are much smaller than the girth of individual parasites (1000–2000 nm). While it is plausible that pore insertion destabilizes membranes sufficiently to allow traversal via gliding motility, PFPs might also work in collaboration with other membrane active proteins that enlarge the pores or further weaken the membrane integrity. Potentially consistent with such collaboration, secreted phospholipases contribute to *Toxoplasma* egress [51] and *Plasmodium* cell traversal [52] and egress [53]. Regardless, much is yet to be learned about the molecular underpinnings of PFP function in the disrupting of biological membranes by apicomplexan parasites.

Acknowledgments: We thank My-Hang (Mae) Huynh and Aric J. Schultz for helpful feedback on this article prior to submission. We also thank Bjorn F. C. Kafsack and Marijo Roiko for help with analyzing ApiPLP sequences and ideas for regulation of pore-forming activity, respectively. Research on PFPs in the Carruthers lab is supported by a grant from the US National Institutes of Health (R01AI120607).

Author Contributions: A.J.G. and V.B.C. wrote the paper and created illustrations for the figures.

Conflicts of Interest: The authors declare no conflict of interest.

References

1. Davies, A.P.; Chalmers, R.M. Cryptosporidiosis. *BMJ* **2009**, *339*, b4168. [CrossRef] [PubMed]
2. Rich, S.M.; Leendertz, F.H.; Xu, G.; LeBreton, M.; Djoko, C.F.; Aminake, M.N.; Takang, E.E.; Diffo, J.L.; Pike, B.L.; Rosenthal, B.M.; et al. The Origin of Malignant Malaria. *Proc. Natl. Acad. Sci. USA* **2009**, *106*, 14902–14907. [CrossRef] [PubMed]
3. Halonen, S.K.; Weiss, L.M. Toxoplasmosis. *Handb. Clin. Neurol.* **2013**, *114*, 125–145. [PubMed]
4. Foreyt, W.J. Coccidiosis and Cryptosporidiosis in Sheep and Goats. *Vet. Clin. N. Am. Food Anim. Pract.* **1990**, *6*, 655–670. [CrossRef]
5. Anderson, M.; Barr, B.; Rowe, J.; Conrad, P. Neosporosis in Dairy Cattle. *Jpn. J. Vet. Res.* **2012**, *60*, S51–S54. [PubMed]
6. Kim, K.; Weiss, L.M. *Toxoplasma gondii*: The Model Apicomplexan. *Int. J. Parasitol.* **2004**, *34*, 423–432. [CrossRef] [PubMed]

7. Dal Peraro, M.; van der Goot, F.G. Pore-Forming Toxins: Ancient, but Never really Out of Fashion. *Nat. Rev. Microbiol.* **2016**, *14*, 77–92. [CrossRef] [PubMed]
8. Dunstone, M.A.; Tweten, R.K. Packing a Punch: The Mechanism of Pore Formation by Cholesterol Dependent Cytolysins and Membrane Attack complex/perforin-Like Proteins. *Curr. Opin. Struct. Biol.* **2012**, *22*, 342–349. [CrossRef] [PubMed]
9. Johnson, T.K.; Henstridge, M.A.; Warr, C.G. MACPF/CDC Proteins in Development: Insights from *Drosophila* Torso-Like. *Semin. Cell Dev. Biol.* **2017**. [CrossRef] [PubMed]
10. Ni, T.; Gilbert, R.J.C. Repurposing a Pore: Highly Conserved Perforin-Like Proteins with Alternative Mechanisms. *Philos. Trans. R. Soc. Lond. B. Biol. Sci.* **2017**, *372*. [CrossRef] [PubMed]
11. Mota, M.M.; Pradel, G.; Vanderberg, J.P.; Hafalla, J.C.; Frevert, U.; Nussenzweig, R.S.; Nussenzweig, V.; Rodriguez, A. Migration of *Plasmodium* Sporozoites through Cells before Infection. *Science* **2001**, *291*, 141–144. [CrossRef] [PubMed]
12. Amino, R.; Giovannini, D.; Thiberge, S.; Gueirard, P.; Boisson, B.; Dubremetz, J.F.; Prevost, M.C.; Ishino, T.; Yuda, M.; Menard, R. Host Cell Traversal is Important for Progression of the Malaria Parasite through the Dermis to the Liver. *Cell Host Microbe* **2008**, *3*, 88–96. [CrossRef] [PubMed]
13. Risco-Castillo, V.; Topcu, S.; Marinach, C.; Manzoni, G.; Bigorgne, A.E.; Briquet, S.; Baudin, X.; Lebrun, M.; Dubremetz, J.F.; Silvie, O. Malaria Sporozoites Traverse Host Cells within Transient Vacuoles. *Cell Host Microbe* **2015**, *18*, 593–603. [CrossRef] [PubMed]
14. Soldati, D.; Meissner, M. *Toxoplasma* as a Novel System for Motility. *Curr. Opin. Cell Biol.* **2004**, *16*, 32–40. [CrossRef] [PubMed]
15. Kafsack, B.F.; Pena, J.D.; Coppens, I.; Ravindran, S.; Boothroyd, J.C.; Carruthers, V.B. Rapid Membrane Disruption by a Perforin-Like Protein Facilitates Parasite Exit from Host Cells. *Science* **2009**, *323*, 530–533. [CrossRef] [PubMed]
16. Acharya, P.; Garg, M.; Kumar, P.; Munjal, A.; Raja, K.D. Host-Parasite Interactions in Human Malaria: Clinical Implications of Basic Research. *Front. Microbiol.* **2017**, *8*, 889. [CrossRef] [PubMed]
17. Robert-Gangneux, F.; Darde, M.L. Epidemiology of and Diagnostic Strategies for Toxoplasmosis. *Clin. Microbiol. Rev.* **2012**, *25*, 264–296. [CrossRef] [PubMed]
18. Meis, J.F.; Verhave, J.P.; Jap, P.H.; Meuwissen, J.H. An Ultrastructural Study on the Role of Kupffer Cells in the Process of Infection by *Plasmodium berghei* Sporozoites in Rats. *Parasitology* **1983**, *86*, 231–242. [CrossRef] [PubMed]
19. Kafsack, B.F.C.; Carruthers, V.B. Apicomplexan Perforin-Like Proteins. *Commun. Integr. Biol.* **2010**, *3*, 18–23. [CrossRef] [PubMed]
20. Ishino, T.; Yano, K.; Chinzei, Y.; Yuda, M. Cell-Passage Activity is Required for the Malarial Parasite to Cross the Liver Sinusoidal Cell Layer. *PLoS Biol.* **2004**, *2*, e4. [CrossRef] [PubMed]
21. Yang, A.S.; O'Neill, M.T.; Jennison, C.; Lopaticki, S.; Allison, C.C.; Armistead, J.S.; Erickson, S.M.; Rogers, K.L.; Ellisdon, A.M.; Whisstock, J.C.; et al. Cell Traversal Activity is Important for *Plasmodium falciparum* Liver Infection in Humanized Mice. *Cell Rep.* **2017**, *18*, 3105–3116. [CrossRef] [PubMed]
22. Hamaoka, B.Y.; Ghosh, P. Structure of the Essential *Plasmodium* Host Cell Traversal Protein SPECT1. *PLoS ONE* **2014**, *9*, e114685. [CrossRef] [PubMed]
23. Holm, L.; Rosenstrom, P. Dali Server: Conservation Mapping in 3D. *Nucleic Acids Res.* **2010**, *38*, W545–W549. [CrossRef] [PubMed]
24. Brett, T.J.; Legendre-Guillemin, V.; McPherson, P.S.; Fremont, D.H. Structural Definition of the F-Actin-Binding THATCH Domain from HIP1R. *Nat. Struct. Mol. Biol.* **2006**, *13*, 121–130. [CrossRef] [PubMed]
25. Hao, W.; Collier, S.M.; Moffett, P.; Chai, J. Structural Basis for the Interaction between the Potato Virus X Resistance Protein (Rx) and its Cofactor Ran GTPase-Activating Protein 2 (RanGAP2). *J. Biol. Chem.* **2013**, *288*, 35868–35876. [CrossRef] [PubMed]
26. Zhao, J.; Beyrakhova, K.; Liu, Y.; Alvarez, C.P.; Bueler, S.A.; Xu, L.; Xu, C.; Boniecki, M.T.; Kanelis, V.; Luo, Z.Q.; et al. Molecular Basis for the Binding and Modulation of V-ATPase by a Bacterial Effector Protein. *PLoS Pathog.* **2017**, *13*, e1006394. [CrossRef] [PubMed]
27. Yue, P.; Zhang, Y.; Mei, K.; Wang, S.; Lesigang, J.; Zhu, Y.; Dong, G.; Guo, W. Sec3 Promotes the Initial Binary t-SNARE Complex Assembly and Membrane Fusion. *Nat. Commun.* **2017**, *8*, 14236. [CrossRef] [PubMed]

28. Nishimura, K.; Han, L.; Bianchi, E.; Wright, G.J.; de Sanctis, D.; Jovine, L. The Structure of Sperm Izumo1 Reveals Unexpected Similarities with Plasmodium Invasion Proteins. *Curr. Biol.* **2016**, *26*, R661–R662. [CrossRef] [PubMed]

29. Ohto, U.; Ishida, H.; Krayukhina, E.; Uchiyama, S.; Inoue, N.; Shimizu, T. Structure of IZUMO1-JUNO Reveals Sperm-Oocyte Recognition during Mammalian Fertilization. *Nature* **2016**, *534*, 566–569. [CrossRef] [PubMed]

30. Kariu, T.; Ishino, T.; Yano, K.; Chinzei, Y.; Yuda, M. CelTOS, a Novel Malarial Protein that Mediates Transmission to Mosquito and Vertebrate Hosts. *Mol. Microbiol.* **2006**, *59*, 1369–1379. [CrossRef] [PubMed]

31. Alves, E.; Salman, A.M.; Leoratti, F.; Lopez-Camacho, C.; Viveros-Sandoval, M.E.; Lall, A.; El-Turabi, A.; Bachmann, M.F.; Hill, A.V.; Janse, C.J.; et al. Evaluation of Plasmodium Vivax Cell-Traversal Protein for Ookinetes and Sporozoites as a Preerythrocytic P. Vivax Vaccine. *Clin. Vaccine Immunol.* **2017**, *24*. [CrossRef] [PubMed]

32. Espinosa, D.A.; Vega-Rodriguez, J.; Flores-Garcia, Y.; Noe, A.R.; Munoz, C.; Coleman, R.; Bruck, T.; Haney, K.; Stevens, A.; Retallack, D.; et al. The *Plasmodium falciparum* Cell-Traversal Protein for Ookinetes and Sporozoites as a Candidate for Preerythrocytic and Transmission-Blocking Vaccines. *Infect. Immun.* **2017**, *85*. [CrossRef] [PubMed]

33. Jimah, J.R.; Salinas, N.D.; Sala-Rabanal, M.; Jones, N.G.; Sibley, L.D.; Nichols, C.G.; Schlesinger, P.H.; Tolia, N.H. Malaria Parasite CelTOS Targets the Inner Leaflet of Cell Membranes for Pore-Dependent Disruption. *Elife* **2016**, *5*. [CrossRef] [PubMed]

34. Tavares, J.; Amino, R.; Menard, R. The Role of MACPF Proteins in the Biology of Malaria and Other Apicomplexan Parasites. *Subcell. Biochem.* **2014**, *80*, 241–253. [PubMed]

35. Lukoyanova, N.; Hoogenboom, B.W.; Saibil, H.R. The Membrane Attack Complex, Perforin and Cholesterol-Dependent Cytolysin Superfamily of Pore-Forming Proteins. *J. Cell. Sci.* **2016**, *129*, 2125–2133. [CrossRef] [PubMed]

36. Garg, S.; Sharma, V.; Ramu, D.; Singh, S. In Silico Analysis of Calcium Binding Pocket of Perforin Like Protein 1: Insights into the Regulation of Pore Formation. *Syst. Synth. Biol.* **2015**, *9*, 17–21. [CrossRef] [PubMed]

37. Roiko, M.S.; Carruthers, V.B. Functional Dissection of *Toxoplasma gondii* Perforin-Like Protein 1 Reveals a Dual Domain Mode of Membrane Binding for Cytolysis and Parasite Egress. *J. Biol. Chem.* **2013**, *288*, 8712–8725. [CrossRef] [PubMed]

38. Roiko, M.S.; Svezhova, N.; Carruthers, V.B. Acidification Activates *Toxoplasma gondii* Motility and Egress by Enhancing Protein Secretion and Cytolytic Activity. *PLoS Pathog.* **2014**, *10*, e1004488. [CrossRef] [PubMed]

39. Garg, S.; Agarwal, S.; Kumar, S.; Shams Yazdani, S.; Chitnis, C.E.; Singh, S. Calcium-Dependent Permeabilization of Erythrocytes by a Perforin-Like Protein during Egress of Malaria Parasites. *Nat. Commun.* **2013**, *4*, 1736. [CrossRef] [PubMed]

40. Law, R.H.; Lukoyanova, N.; Voskoboinik, I.; Caradoc-Davies, T.T.; Baran, K.; Dunstone, M.A.; D'Angelo, M.E.; Orlova, E.V.; Coulibaly, F.; Verschoor, S.; et al. The Structural Basis for Membrane Binding and Pore Formation by Lymphocyte Perforin. *Nature* **2010**, *468*, 447–451. [CrossRef] [PubMed]

41. Yang, A.S.; Boddey, J.A. Molecular Mechanisms of Host Cell Traversal by Malaria Sporozoites. *Int. J. Parasitol.* **2017**, *47*, 129–136. [CrossRef] [PubMed]

42. Ishino, T.; Chinzei, Y.; Yuda, M. A *Plasmodium* Sporozoite Protein with a Membrane Attack Complex Domain is Required for Breaching the Liver Sinusoidal Cell Layer Prior to Hepatocyte Infection. *Cell. Microbiol.* **2005**, *7*, 199–208. [CrossRef] [PubMed]

43. Tavares, J.; Formaglio, P.; Thiberge, S.; Mordelet, E.; Van Rooijen, N.; Medvinsky, A.; Menard, R.; Amino, R. Role of Host Cell Traversal by the Malaria Sporozoite during Liver Infection. *J. Exp. Med.* **2013**, *210*, 905–915. [CrossRef] [PubMed]

44. Pradel, G.; Frevert, U. Malaria Sporozoites Actively Enter and Pass through Rat Kupffer Cells Prior to Hepatocyte Invasion. *Hepatology* **2001**, *33*, 1154–1165. [CrossRef] [PubMed]

45. Deligianni, E.; Morgan, R.N.; Bertuccini, L.; Wirth, C.C.; Silmon de Monerri, N.C.; Spanos, L.; Blackman, M.J.; Louis, C.; Pradel, G.; Siden-Kiamos, I. A Perforin-Like Protein Mediates Disruption of the Erythrocyte Membrane during Egress of *Plasmodium berghei* Male Gametocytes. *Cell. Microbiol.* **2013**, *15*, 1438–1455. [CrossRef] [PubMed]

46. Kadota, K.; Ishino, T.; Matsuyama, T.; Chinzei, Y.; Yuda, M. Essential Role of Membrane-Attack Protein in Malarial Transmission to Mosquito Host. *Proc. Natl. Acad. Sci. USA* **2004**, *101*, 16310–16315. [CrossRef] [PubMed]

47. Ecker, A.; Bushell, E.S.; Tewari, R.; Sinden, R.E. Reverse Genetics Screen Identifies Six Proteins Important for Malaria Development in the Mosquito. *Mol. Microbiol.* **2008**, *70*, 209–220. [CrossRef] [PubMed]

48. Wade, K.R.; Tweten, R.K. The Apicomplexan CDC/MACPF-Like Pore-Forming Proteins. *Curr. Opin. Microbiol.* **2015**, *26*, 48–52. [CrossRef] [PubMed]

49. Lopez, J.A.; Brennan, A.J.; Whisstock, J.C.; Voskoboinik, I.; Trapani, J.A. Protecting a Serial Killer: Pathways for Perforin Trafficking and Self-Defense Ensure Sequential Target Cell Death. *Trends Immunol.* **2012**, *33*, 406–412. [CrossRef] [PubMed]

50. Brennan, A.J.; Chia, J.; Browne, K.A.; Ciccone, A.; Ellis, S.; Lopez, J.A.; Susanto, O.; Verschoor, S.; Yagita, H.; Whisstock, J.C.; et al. Protection from Endogenous Perforin: Glycans and the C Terminus Regulate Exocytic Trafficking in Cytotoxic Lymphocytes. *Immunity* **2011**, *34*, 879–892. [CrossRef] [PubMed]

51. Pszenny, V.; Ehrenman, K.; Romano, J.D.; Kennard, A.; Schultz, A.; Roos, D.S.; Grigg, M.E.; Carruthers, V.B.; Coppens, I. A Lipolytic Lecithin:Cholesterol Acyltransferase Secreted by Toxoplasma Facilitates Parasite Replication and Egress. *J. Biol. Chem.* **2016**, *291*, 3725–3746. [CrossRef] [PubMed]

52. Bhanot, P.; Schauer, K.; Coppens, I.; Nussenzweig, V. A Surface Phospholipase is Involved in the Migration of *Plasmodium* Sporozoites through Cells. *J. Biol. Chem.* **2005**, *280*, 6752–6760. [CrossRef] [PubMed]

53. Burda, P.C.; Roelli, M.A.; Schaffner, M.; Khan, S.M.; Janse, C.J.; Heussler, V.T. A *Plasmodium* Phospholipase is Involved in Disruption of the Liver Stage Parasitophorous Vacuole Membrane. *PLoS Pathog.* **2015**, *11*, e1004760. [CrossRef] [PubMed]

toxins

MDPI

Article

Interaction of Cholesterol with Perfringolysin O: What Have We Learned from Functional Analysis?

Sergey N. Savinov and Alejandro P. Heuck *

Department of Biochemistry and Molecular Biology, University of Massachusetts, Amherst, MA 01003, USA; ssavinov@umass.edu
* Correspondence: heuck@umass.edu; Tel.: +1-413-545-2497

Academic Editor: Alexey S. Ladokhin
Received: 31 October 2017; Accepted: 17 November 2017; Published: 23 November 2017

Abstract: Cholesterol-dependent cytolysins (CDCs) constitute a family of pore-forming toxins secreted by Gram-positive bacteria. These toxins form transmembrane pores by inserting a large β-barrel into cholesterol-containing membranes. Cholesterol is absolutely required for pore-formation. For most CDCs, binding to cholesterol triggers conformational changes that lead to oligomerization and end in pore-formation. Perfringolysin O (PFO), secreted by *Clostridium perfringens*, is the prototype for the CDCs. The molecular mechanisms by which cholesterol regulates the cytolytic activity of the CDCs are not fully understood. In particular, the location of the binding site for cholesterol has remained elusive. We have summarized here the current body of knowledge on the CDCs-cholesterol interaction, with focus on PFO. We have employed sterols in aqueous solution to identify structural elements in the cholesterol molecule that are critical for its interaction with PFO. In the absence of high-resolution structural information, site-directed mutagenesis data combined with binding studies performed with different sterols, and molecular modeling are beginning to shed light on this interaction.

Keywords: cholesterol-dependent cytolysins; Perfringolysin O; cholesterol; cholesterol-binding

1. Introduction

Perfringolysin O (PFO) is the prototypical example of a family of Gram-positive bacterial pore-forming toxins known as the cholesterol-dependent cytolysins (CDCs) [1–3]. Despite being present in a broad range of species, most CDCs show an amino acid sequence identity greater than 39% when compared to PFO [2]. The C-terminus (domain 4 or D4) of PFO is responsible for the cholesterol-dependent membrane binding and is the domain with the highest percentage of amino acid identity among CDC members. Cholesterol recognition via D4 is a distinguishing feature of the CDCs. An exception was found for intermedilysin because it uses the human receptor CD59 as a receptor for membrane targeting [4]. However, intermedilysin still requires cholesterol to form pores in membranes [5].

It has long been known that a high level of cholesterol is required in membranes to trigger PFO binding [6–8]. More recently it was shown that how much cholesterol is required to trigger binding depends on the overall lipid composition of the membrane [9,10]. However, the precise mechanism by which cholesterol triggers binding and the conformational changes that lead to pore-formation are unknown. In this work we will review our current knowledge on CDC-cholesterol interaction and present some additional insights on the interaction between cholesterol and PFO.

1.1. Structural Elements of Domain 4 Involved in Cholesterol Recognition

PFO D4 consists of two four-stranded β-sheets located at the C-terminus of the protein (Figure 1A). There are four loops that interconnect the eight β-strands at the distal tip of the toxin. These loops insert into the membrane upon binding and are presumably responsible for the interaction of the toxin with cholesterol [11–13]. Two of these loops (L2 and L3) connect β-strands from opposite β-sheets, while L1 and the undecapeptide connect β-strands from the same β-sheet. L1 and the undecapeptide are parallel to each other and abutted perpendicularly by L2, forming a pocket in the bottom of the protein (Figure 1B). The loops that form this pocket are the most conserved segments in D4, and modifications to any of these three loops affect the cholesterol binding properties of PFO [13–16]. The remaining L3 is less conserved (Figure 1C). Interestingly, a similar loop arrangement has been recently described for the C-edge loops of the eukaryotic protein arrestin [17], a protein that interacts with G protein-coupled receptors blocking G-protein-mediated signaling and directs the receptors for internalization.

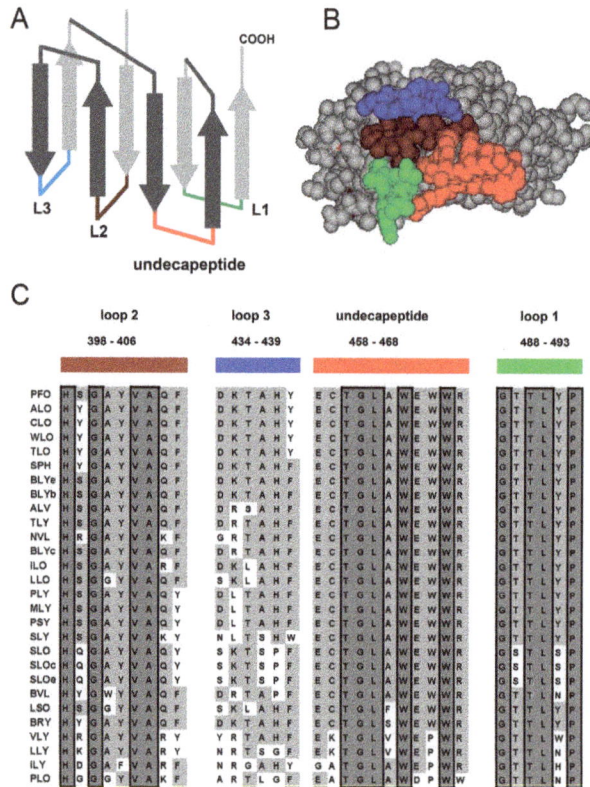

Figure 1. The loops at the tip of D4 are highly conserved among CDCs. (**A**) Cartoon representation of the PFO D4 β-sandwich showing the location of the loops and the conserved undecapeptide. The undecapeptide was colored red and the loops were colored green (L1), brown (L2), and blue (L3); (**B**) A view of the tip of PFO D4 from the bottom showing the loops and undecapeptide color coded as in A; (**C**) Sequence alignment of the 28 CDC family members showing the conserved amino acids boxed and with dark grey background. Highly conserved amino acids are shown with a light grey background. Protein names were abbreviated as defined in [2].

The undecapeptide is the longest and most conserved of the four loops. It was originally thought to be exclusively responsible for cholesterol recognition and binding. This idea was supported by several studies showing that modifications in it greatly decreased the pore-forming activity of the protein [14,18–23]. However, more recent studies showed that other loops in D4 are also responsible for cholesterol recognition [13]. The undecapeptide has now been suggested to play a role in both the pre-pore to pore transition [12] and the coupling of monomer binding with initiation of the pre-pore assembly [24]. Dowd and colleagues recently showed that modification of a charged amino acid in the undecapeptide (R468) resulted in complete elimination of the pore-forming activity of PFO and had a significant effect on the membrane binding of the toxin [14,24]. Despite the novel functions assigned to the undecapeptide, its role in binding cannot be neglected since many modifications to this segment have been shown to have a significant effect in toxin-membrane interaction [14].

The L3 is located on the far edge of D4, away from a nascent cavity formed by the undecapeptide, L1, and L2 (Figure 1). Modifications introduced into L3 have been shown either to have a negligible effect on cholesterol interaction, or to decrease the amount of cholesterol required for binding [13,16,25]. These results suggest that L3 plays a limited role in cholesterol recognition, and its effect on binding may be related to nonspecific interactions with the membrane that stabilize the bound form of the monomer (e.g., decreasing the k_{off}).

A suggested cholesterol recognition motif composed by only two adjacent amino acids in L1, (T490 and L491 in PFO, Figure 1C) [13], is conserved throughout all reported CDCs. Modifications to these two amino acids greatly affect the binding of the protein to both cell and model membranes [13,26]. These data suggest a prominent role for T490 and L491 in cholesterol recognition, however, other well conserved amino acids located in proximity of the pocket formed by L1, L2, and the undecapeptide have not been analyzed yet (e.g., H398, Y402, A404, E458, and P493) and may also contribute to cholesterol binding.

1.2. The Effects of Membrane Lipids on the Cholesterol Threshold Required for CDC Binding

Cholesterol concentrations of more than 30 mol % are usually required to trigger binding of PFO to liposomes prepared exclusively with phosphatidylcholine [8,27]. Other CDCs showed similar effects, for example streptolysin O (SLO) [7], lysteriolysin O [28], and tetanolysin [6]. How much cholesterol is required to trigger PFO binding (or "cholesterol threshold") is reduced by the incorporation of double bonds in the acyl chains of the phospholipids or by replacing phosphatidylcholine by phospholipids with smaller head groups [9,10,15]. The high level of cholesterol required to trigger PFO binding, the discovery of cholesterol-rich domains in membranes, and the presence of PFO on detergent resistant membranes [29] led some researchers to associate PFO binding with the presence of membrane rafts [30]. However, it is difficult to envision a scenario where cholesterol will be more readily available to interact with PFO if located in a cholesterol-rich domain where the interaction with other lipids is stronger. For example, it has been shown that the presence of sphingomyelin (a lipid that interacts with cholesterol) actually interferes with PFO binding [10]. Recent studies on PFO-cholesterol interaction suggest that accessibility of cholesterol to the membrane surface is the key factor to trigger PFO binding [10,15,31,32].

Moreover, despite the influence phospholipids have on the cholesterol-dependent binding of PFO, their presence is not required since cholesterol alone (in the absence of any other lipid) is sufficient to trigger PFO oligomerization and formation of ring-like complexes ([33] and references therein).

1.3. Structure Elements of Cholesterol that Influence CDC Activity

Early studies of the inhibition of SLO and PFO hemolytic activity by different sterols revealed elements of the cholesterol molecule that are critical for its interaction with the CDCs [34–36]. The affinity of the toxin for a particular sterol was indirectly estimated by measuring the hemolytic activity of the toxin after a pre-incubation with the sterol. It was assumed that the higher the inhibition, the stronger the affinity for the sterol (Table 1, Figure 2). Results from these studies have been reviewed by Alouf [37] and are briefly summarized below.

Table 1. Interaction of different sterols with CDCs.

	Hemolysis Inhibition		Sterol in Aqueous Buffer		Sterol in Liposomes	
	SLO	PFO	PFO	PFO	PFO	PFO
			Trp D4	NBD D3	Trp D4	Oligo SDS
cholesterol	1.0	1.0	1.0	1.0	1.0	+++
7-dehydrocholesterol	(1.7)	0.67	quenched	0.89	quenched	
dihydrocholesterol	0.50	0.46	0.83		1.1	+++
β-sitosterol	0.50	0.61	0.59	0.80	0.94	+++
lathosterol	0.50				0.85	+++
allocholesterol		0.40			0.68	++
desmosterol		0.18			1.2	+++
coprostanol	0.71	0.11			0.65	+
zymosterol					0.61	+
ergosterol	0.10	0.13	quenched	0.44	quenched	
fucosterol			0.42			
stigmasterol	0.33	0.037	low	<0.08		

Numbers in the table show the relative effect when compared with the one observed for cholesterol. Hemolysis inhibition is calculated using the Inhibitory dose 50 (I_{50}) reported for SLO [34] or PFO [36]. Values for the interaction of PFO with sterols in aqueous buffer were calculated using the concentration of sterol that cause half of the total change in Trp emission (nPFO, Figure 5 and Ref. [33]) or NBD emission (rPFO$^{V322C-NBD}$, Figure 6 and Ref. [33]). Sterol in liposomes values were calculated using the mol % sterol that cause half of the total change in Trp emission for rPFO [9]. Relative values for rPFO oligomerization were estimated using the SDS-agarose gel electrophoresis analysis done by Nelson et al. [9].

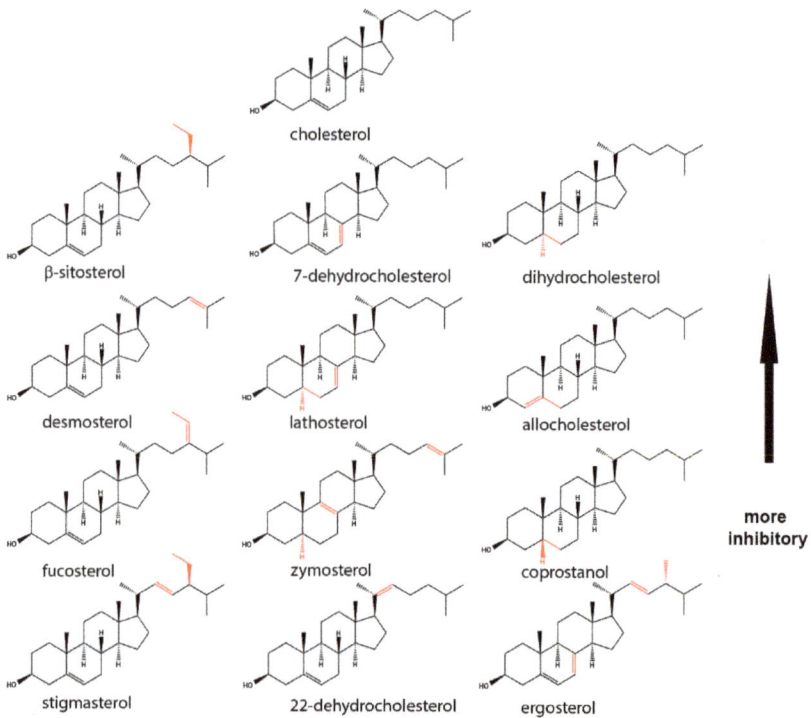

Figure 2. Chemical structure of the sterols that interacts with the CDCs. The differences from cholesterol are highlighted in red. Top molecules inhibit/interact strongly, and the ones in the bottom more weakly (based on data presented on Table 1).

1.3.1. The Presence of a Lateral Aliphatic Side Chain of Suitable Size at Carbon 17 Is Required

Addition of polar hydroxyl groups at position C25, C26, or C20 of the eight-carbon chain removes the inhibitory effect (Figure 3). Replacement of the eight carbon acyl chain for a keto group or an acetyl group removes the inhibitory effect. Sterols with a double bond at C24–C25 (desmosterol) or with a =CH-CH3 group at C24 (fucosterol) are still inhibitory. Modification of the eight-carbon chain by introduction of an ethyl group at C-24 (β-sitosterol) is not critical, but the simultaneous addition of a double bond at C22–C23 and either an ethyl group (stigmasterol) or a methyl group (ergosterol) at C24 weakens the inhibitory effect (see Figure 2).

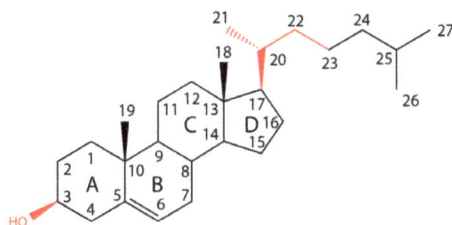

Figure 3. Chemical structure of cholesterol showing the individual rings (**A**)–(**D**) and numbered carbon atoms. Elements identified as critical for the interaction with CDCs are indicated in red.

1.3.2. The Presence of a 3 β-Hydroxy Group on Ring A Is Required

The inhibitory capacity of the sterol is removed when the hydroxyl group is eliminated (cholestane), oxidized (cholestanone), esterified (cholesterol acetate), etherified (cholesterol methyl ether), or epimerized into alpha position (epicholesterol). Substitution of the hydroxyl group for a thiol group (thiocholesterol) or chloride (3 chlorocholestene) also removes the inhibitory effect.

1.3.3. An Intact Ring B Is Required

The presence of the A ring with the β-hydroxyl group and the aliphatic chain at carbon 17 are not sufficient for binding if the B ring is open (cholecalciferol). However, neither the saturated or unsaturated state of ring B and the position of double bonds (lathosterol, allocholesterol, or zymostenol) nor the stereochemical relationships of rings A and B to each other are critical for inhibition. The 5β-cis (coprostanol) and 5α-trans (dihydrocholesterol) configurations are both inhibitory.

Similar effects were observed for the inhibitory effect of sterols on SLO and PFO (the inhibition of 7-dehydrocholesterol was higher for SLO, but the sample used in this study presented 3 spots on a thin layer chromatography plate, therefore we need to be cautious when considering this result). An exception was coprostanol, which was a better inhibitor for SLO than for PFO. Interestingly, some amino acids in L1 and L2 differ between SLO and PFO (Figure 1C), suggesting that these loops may interact with the B ring of cholesterol.

Some differences were observed when the sterols were incorporated into model membranes (Table 1) [9], but in this case one also need to consider the differential interaction that each sterol may have with the phospholipids. Oligomerization of PFO on liposomes containing ergosterol or 7-dehydrocholesterol was similar to the one observed with cholesterol, but these sterols quenched the Trp emission [38] and therefore the binding of PFO to liposomes could not be assessed using Trp fluorescence.

In the present work, we explored the interaction of free sterols in solution with PFO using the Trp emission increase that follows D4-sterol interaction [33]. In addition, we study the effect of sterol-binding in the conformational changes that occur in D3. Finally, molecular modeling was attempted to offer structural rationale for the observed trends.

2. Results

2.1. Selective Solubilization of Sterol Aggregates by Methyl-β-cyclodextrin

We have shown that PFO is able to bind to cholesterol aggregates in solution. Cholesterol aggregates remain soluble in neutral aqueous buffers, but start to precipitate when the cholesterol concentration reaches the solubility limit (around 4.7 μM) [33,39]. Similar aggregation profiles were observed when other sterols were added into aqueous buffer up to a concentration of 30 μM (Figure 4A) with the exception of 22-dehydrocholesterol, where scatter was lower than that observed for the other sterols used in this study. It is well known that cholesterol interacts with methyl-β-cyclodextrin (mCD), and the addition of mCD solubilizes cholesterol aggregates and microcrystals (Figure 4B) [33]. While attempting to repeat the mCD solubilization with other sterols, we noticed that the sole addition of one or two carbons to C24 in the aliphatic chain of the sterols was sufficient to interfere with this process. No solubilization was observed for aggregates formed by β-sitosterol, fucosterol, stigmasterol, or ergosterol (Figures 2 and 4B), but complete solubilization was observed for cholesterol, 7-dehydrocholesterol, and dihydrocholesterol. The scattered light (relative units) for aggregates formed by 30 μM aqueous solutions of two other sterols lacking modifications to the acyl chain -22-dehydrocholesterol and epicholesterol- was 71,800 and 186,400, respectively. The scattered light for both aggregates decreased more than 96% after addition of mCD, in good agreement with the need for a C20–C25 linear aliphatic chain in the sterol molecule for fast mCD solubilization of sterol aggregates. Yet, it was reported that mCD can bind some C24 substituted sterols if they are added directly into the solution containing mCD [40].

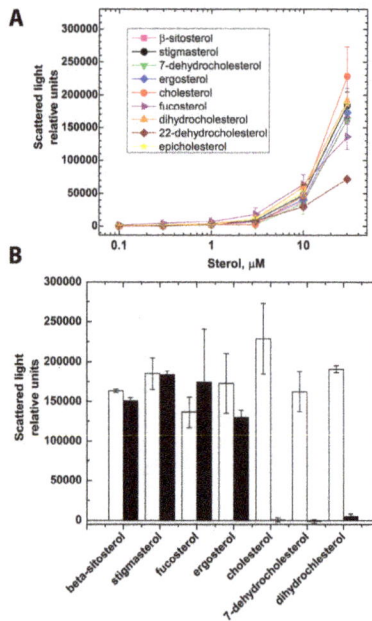

Figure 4. Sterol precipitated when added into aqueous buffer solution and they were differentially solubilized by mCD. (**A**) Scattered light at 500 nm of aqueous buffer solutions containing the indicated amount of sterols. Sterols were added from ethanolic solutions incrementally and incubated 5 min at 37 °C before each measurement; (**B**) mCD (final concentration 3 mM) was added into solutions containing 30 μM sterols and the right angle light scatter measured after 5 min incubation at 37 °C. The bars represent the average of at least two measurements and the error bars correspond to the range. White bars and black bars represent the scattered light before and after incubation with mCD, respectively.

The molecular bases for the sterol-mCD interactions are not well understood, but it is clear from the observations described above that if sterols with group additions to C24 are allowed to form aggregates, they do not interact with mCD in the same way that they do when directly diluted into a solution containing mCD. These results suggest that the order of the addition of the sterols and the protein may influence the outcome obtained for protein-sterol interactions. Therefore, we reasoned that when studying the interaction of sterols with water soluble molecules (like PFO in these studies) it would be necessary to add the sterols into a solution containing PFO to minimize the formation of sterol aggregates, and maximize the exposure of PFO to solubilized sterol monomers.

2.2. PFO Interaction with Free Sterols

Liposomes made with different sterols have also been used to study how modifications to the cholesterol molecule affect its interaction with PFO. However, in these studies the PFO-sterol interaction will be influenced by both, the direct interaction (affinity) of the sterol molecule with PFO, and the interaction of the sterol with other membrane components (phospholipids, sphingomyelin, etc.). The higher the interaction of the sterol with other lipids, the less available it will be to interact with PFO. Therefore, to determine what elements of the cholesterol molecule are critical to bind PFO, it is important to perform these studies in the absence of other lipid components. We have shown that binding of PFO to cholesterol in aqueous solution produces an increase in Trp emission, similar to the one observed when the toxin binds to membranes containing cholesterol [33]. We reasoned that the same emission change could be used to analyze the interaction with other sterols.

The interaction of PFO with sterols was studied following the Trp emission increase that follows PFO-sterol interaction (Figure 5). No emission increase was observed when PFO was incubated with non-interacting sterols like epicholesterol (Figure 5) [33]. In this analysis, both dihydrocholesterol (reduction of the C5–C6 double bond) and β-sitosterol (addition of an iso-propyl group at C24) showed a concentration-dependent Trp emission profile that was slightly shifted to higher sterol concentration when compared to the one obtained for cholesterol (Figure 5). Similarly, the concentration-dependent change when adding fucosterol was shifted to higher sterol concentrations, indicating that the rigidity introduced by the double bond between C24 and the ethyl group restrict the interaction of the sterol acyl tail with PFO (see Figure 2). These four sterols showed a similar maximal increase in Trp emission when added to a final concentration of 10 μM. A lower Trp emission increase was observed when PFO was incubated with stigmasterol or 22-dehydrocholesterol, suggesting that flexibility between C20–C22 is important for the interaction of PFO with cholesterol. Surprisingly, no Trp emission increase was observed for 7-dehydrocholesterol and ergosterol, two sterols that are able to inhibit the hemolytic activity of PFO (Table 1) [36]. Both of these sterols possess two conjugated double bonds in the B ring. It has been suggested that this double bond quenches the Trp emission eliminating the increase produced upon the interaction of PFO with the sterol molecules [41]. Inner filter effect could also contribute to mask the Trp emission increase because of the overlap between the Trp and sterol absorption wavelengths (Figure S1). Therefore, a different approach was required to analyze the interaction of these sterols with PFO.

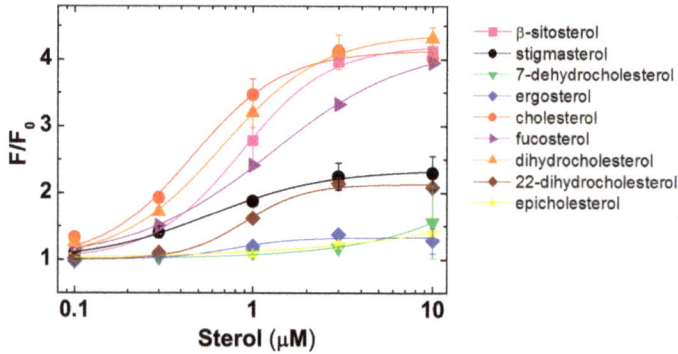

Figure 5. Binding of sterols to the PFO derivative containing the native undecapeptide (nPFO) [33]. Trp emission intensity for 0.1 μM nPFO was measured before (F_0) and after (F) addition of the indicated amount of sterol. Each data point shows the average of at least two measurements and their range. The cholesterol concentration that produced half of the total Trp emission increase for nPFO was 0.5 μM.

The increase in Trp emission that results from the interaction of PFO D4 with cholesterol is followed by the movement of a short β-strand (β5) in D3 that exposes the monomer-monomer interface required for oligomer formation [42]. This conformational change can be detected using the rPFO$^{V322C-NBD}$ derivative. The environment-sensitive NBD fluorescent probe has a high lifetime (~8 ns) in the monomeric toxin, and NBD lifetime drops to ~1 ns when the protein interacts with cholesterol or cholesterol containing membranes [33]. A decrease in the fluorescence intensity of the NBD dye would be indicative of the interaction of PFO with sterols even if the spectroscopic properties of the sterol molecule interfere with the increase of Trp emission (as it is the case for 7-dehydrocholesterol and ergosterol). Using this assay, we observed that both 7-dehydrocholesterol (extra double bond in ring B) and ergosterol (the same extra double bond in ring B plus another double bond at C22 and a methyl group at C24) triggered the NBD emission decrease when incubated with the rPFO$^{V322C-NBD}$ derivative (Figure 6A,C). We also tested β-sitosterol and stigmasterol, two sterols that showed a strong and weak interaction with PFO, respectively (Figure 6B,D). In both cases the decrease in NBD emission was parallel to the increase of Trp emission. These data indicate that the conformational change in D3 is also a good reporter for the interaction of PFO with different sterols.

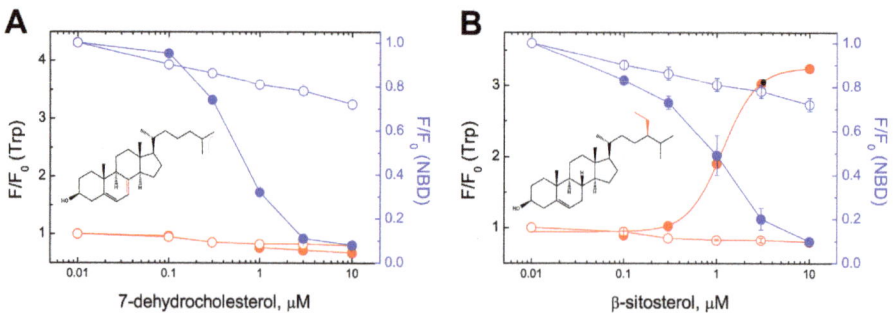

Figure 6. *Cont.*

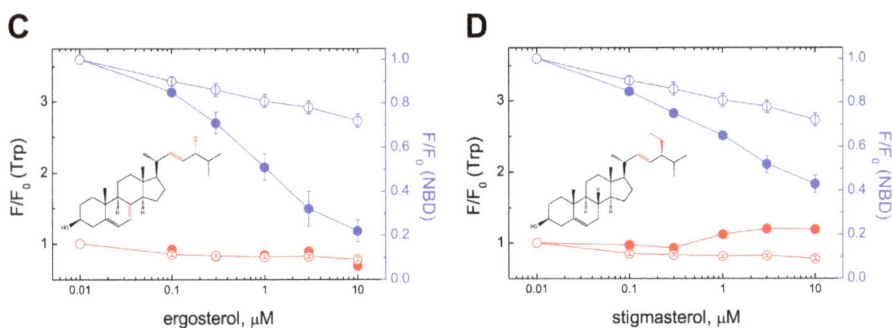

Figure 6. Sterol binding to D4 paralleled the conformational change in D3. rPFO$^{V322C-NBD}$ (0.1 μM) was titrated with (**A**) 7-dehydrocholesterol, (**B**) β-sitosterol, (**C**) ergosterol, (**D**) stigmasterol, and the Trp emission intensity and NBD emission intensity were determined in the same sample at each sterol concentration. Open symbols correspond to a parallel experiment where an identical volume of ethanol was added (no-sterol control). Trp emission data are shown in red, whereas NBD emission data are shown in blue. The average of at least two independent measurements and their range are shown. As a reference, the cholesterol concentration that produced half of the total NBD emission decrease for rPFO$^{V322C-NBD}$ was 0.8 μM [33].

2.3. Molecular Modeling Rationales for the Observed Cholesterol Structure–Activity Relationship and Mutagenesis Data

To rationalize the data from both the site-directed mutagenesis studies and Structure–Activity Relationship (SAR) analysis of sterols, we have undertaken a modeling study using available CDC structures as receptors and cholesterol as a ligand. The realization that conformational changes need to occur to create an arrangement capable of association with cholesterol, prompted us to employ an Induced Fitting Docking (IFD) algorithm [43] (Schrödinger, LLC, New York, NY, USA). The IFD algorithm iterates docking stages with local minimizations to identify a likely binding site on the surface of D4. We have selected the following criteria for judging the likelihood of putative binding arrangements: (i) contact with highly conserved residues of D4 (due to mechanistic similarities in cholesterol effects among CDCs); (ii) engagement of both the conserved Thr-Leu pair on L1 and undecapeptide (as determined through site-directed mutagenesis); and (iii) participation of equatorial hydroxyl, the only polar site in cholesterol, in H-bonding interaction with conserved donors, acceptors, or both.

Our initial attempts with the PFO structures available from Protein Data Bank (PDB, ID: 1PFO, 1M3I, and 1M3J) failed to yield any reasonable binding arrangements that would satisfy the criteria stated above. Further analysis of these and other structures of CDCs highlighted the unique conformation that the undecapeptide adopts in PFO [26,44,45], which is curled up against the exposed face of the β-sandwich. The undecapeptide is kept in this conformation by an edge-to-face stacking of conserved W464 with Y432, a residue that is unique to PFO among the various solved CDCs x-ray structures. This arrangement separates the undecapeptide from L1 with its Thr-Leu pair, critical for cholesterol association [13], and was, therefore, not as suitable as a starting point for these studies without undergoing a major conformational reorganization.

As a consequence, we have expanded the search to suitable structural models for docking studies to all CDCs featuring the intact undecapeptide sequence from PFO. This search yielded eleven X-ray crystal structures from the PDB with resolutions of at least 3.1 Å: anthrolysin O (1 structure), listeriolysin O (1), SLO (1), suilysin (1), and pneumolysin (PLY) (7). Unlike the PFO structures, the seven structures of PLY offered a rather wide diversity of conformational solutions for the undecapeptide (Figure 7). PLY has a high level of conservation with PFO in the membrane-associating regions (91%

identity, 97% similarity for the loops in D4, Figure S2). The conformations of the undecapeptide in PLY ranged from a nearly canonical β-hairpin with $_{430}$GLAW$_{433}$ reverse turn projected away from L1 (PDB ID: 5CR8) to a significantly more unstructured and relatively unraveled loop with multiple solvent-exposed peptide bonds (PDB ID: 4ZGH) that increases the density of hydrophobic residues co-projected toward the membrane (Figure 7). This flexibility is not unexpected for a sequence that contains amino acids uncommon in turns (Leu, Ala, Trp) and strands (Glu, Cys, Gly) [46]. Gratifyingly, in the latter structure, a largely contiguous hydrophobic pocket at the interface of the L1 and undecapeptide is starting to emerge furnished exclusively with conserved residues, several of which come from the rearranged undecapeptide.

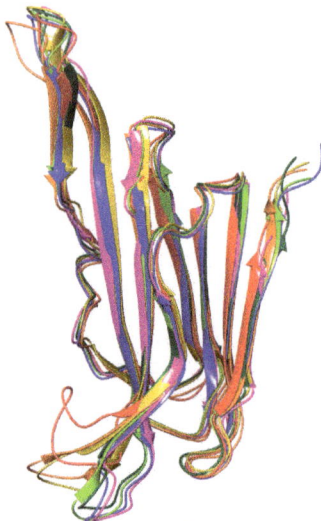

Figure 7. Overlay of PFO (red) and PLY (orange through violet) structures from the PDB with variable level of loop unravelling by the conserved undecapeptide motif (1PFO: red, 5CR8 chain D: orange, 5CR8 chain A: yellow, 5CR6: yellow-green, 5AOD: green, 5AOF: blue, and 4ZGH: violet). The PLY structure with the most proximal disposition of undecapeptide and L1 loop (4ZGH) was used as a starting point for flexible docking of cholesterol.

Upon flexible IFD, the model derived from 4ZGH has yielded a set of related binding poses that were compatible with binding arrangement criteria, cholesterol SAR, and mutagenesis data. Contrary to the previously published models [14,26], the binding pose predicts that cholesterol undergoes a flip from the membrane arrangement (Figure 8A), which places the equatorial hydroxyl of cholesterol within the undecapeptide in multiple H-bonding contacts with the backbone amides (Figure 8B), providing an explanation for the rigorous H-bonding-enable hydroxyl requirement in cholesterol variants capable of hemolysis inhibition (Table 1). This flip could be coupled to the overall inward movement of the undecapeptide loop as it approaches L1 from the curled away positions. In the model, the B-ring plays a role of a rigid spacer between the A ring and C & D rings engaging the conserved L1 residues. Hence, while its rigid cyclic nature is critical for the display of important recognition elements., (e.g., cholecalciferol) [34], the saturation status and even bridgehead stereochemistry are not expected to have a significant effect on binding, as seen with lathosterol, allocholesterol, zymostenol, coprostanol and dihydrocholesterol. The remainder of the polycyclic core of the sterol (rings C and D, Figure 3) is found in contact with the conserved Thr-Leu pair (T490-L491 in PFO or T459-L460 in PLY) in L1, proven essential for binding of cholesterol in previous studies [13]. The aliphatic tail, in turn, interdigitates with similarly aliphatic and conserved V439 and P462 in PLY (V470 and P493 in PFO).

The extended binding pose adopted by cholesterol explains the need for the equatorially projected hydroxyl, which is capable of making multiple H-bonding contacts, unlike a largely occluded axial hydroxyl in epicholesterol [33]. Two other notable events occur during the IFD search to permit the association: (i) the terminal residue of the undecapeptide (R437 in PLY and R468 in PFO), a residue critical for the stability of the toxin structure [47], becomes solvent exposed and opens up the space between two β-sheets to become available for association with cholesterol; and (ii) the first residue of the undecapeptide, (E427 in PLY and E458 in PFO), having lost its salt-bridge partner while remaining in a largely hydrophobic environment is expected to become protonated with a concomitant pKa shift from ~4.5 in PFO or β-hairpin PLY structures to 6.5 and 7.5 in cholesterol-free and bound 4ZGH-derived models, respectively (Epik, Schrödinger, LCC, New York, NY, USA) [48], making the binding surface even more hydrophobic and therefore, more welcoming to a ligand as lipophilic as cholesterol. This observation is consistent with the report of low pH (5.5–6) enhancing PFO–membrane association [9]. This model provides a valuable set of testable hypotheses for further evolution of our insight into mechanistically complex events occurring prior to pore formation by CDCs.

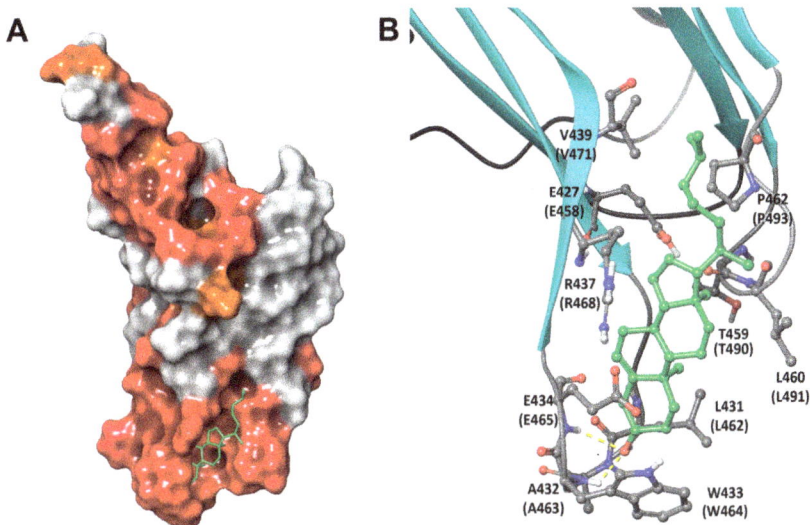

Figure 8. The IFD-proposed model for the cholesterol–PLY-D4 complex. (**A**) PLY D4 is shown in its full size and rendered as molecular surface colored by sequence conservation between PFO and PLY (red: identical, orange: similar, white: non-conserved). The molecule of bound cholesterol is shown as green tubes; (**B**) Close-up of the binding pose predicted by the IFD docking. The binding site of PLY D4 is rendered as ribbons shown as cyan arrows and grey tubes for β-strands and loops, respectively. The key contact residues are shown and labeled with sequence positions for PLY and PFO (in parenthesis). The cholesterol is shown as a green ball-and-stick model, and H-bonds between the hydroxyl of cholesterol and undecapeptide backbone are shown as yellow dashed lines.

3. Discussion

Despite the various X-ray high resolution structures available for CDCs, the structure of the toxin bound to cholesterol remained elusive. The only information available about the interaction of CDCs with cholesterol has been obtained by combining effects of site directed mutagenesis and structurally distinct sterols. Site directed mutagenesis provided information about how different amino acids contribute to cholesterol binding (assuming that amino acid substitutions do not significantly affect the folding of the protein). The use of different sterols showed how critical different groups or parts of the molecule are for protein interaction. In this work we reviewed the information collected on CDC-sterol interactions and provided some further information about the interaction of free sterols with PFO, a prototypical example of the CDCs. By integrating all the available data, we constructed a binding model for cholesterol–D4 complex that satisfies both the previous findings and those reported herein, and rationalizes the critical need for cholesterol in the membrane-anchoring mechanism of CDCs.

Cholesterol and related sterols (Figure 2) are hydrophobic molecules with very low solubility in water. These sterols precipitate when their concentration increases above 5 μM (Figure 4A). Cyclic sugars, like mCD have been often used to solubilize and transport cholesterol into/from membranes [40]. While most aggregated sterols were readily solubilized by mCD, those with bulky substitutions into the acyl-tail were not (Figure 4B). It is not clear if sterols with additions to their acyl chain are not able to interact with mCD (steric effect) or if the formed sterol aggregates are kinetically trapped in a meta-stable state. Successful binding of mCD to some of the C24 substituted sterols suggest that the latter may be the case [40]. More research is required in this area to elucidate the molecular details for the interaction of sterols with mCD. However, to minimize the potential problems of working with aggregated sterols in these studies, each sterol was added in small aliquots into a solution containing monomeric PFO.

Interaction of PFO with sterols was determined by the increase in Trp emission (Figure 5). Cholesterol, dihydrocholesterol, and β-sitosterol showed a similar profile, which was in agreement with the results obtained using inhibition of hemolysis (Table 1). Stigmasterol, a poor inhibitor of hemolysis, showed a small change in the Trp emission. This weak interaction was corroborated by the small decrease on NBD emission in D3, a conformational change that follows binding (Figure 6D). A similar low interaction profile was obtained for 22-dehydrocholesterol, suggesting that free movement around C22 is required to stabilize cholesterol in its binding site. The limited NBD emission decrease observed for ergosterol suggested an intermediate binding affinity for PFO D4 (similar to that of fucosterol, as determined using Trp emission). Flexibility of the acyl chain seems to be necessary to accommodate the cholesterol molecule into the D4 binding site. NBD emission decrease observed for 7-dehydrocholesterol and cholesterol confirmed the similar interaction observed using Trp emission. In summary, two regions of cholesterol appeared to be critical for the interaction with PFO, the β-hydroxyl group and the flexibility of the acyl chain around C20–C22 (Figure 3).

The extent of the decrease in NBD emission observed for 7-dehydrocholesterol, β-sitosterol, ergosterol, stigmasterol, and the non-interacting epicholesterol correlated with the inhibitory properties of the same sterols when tested using hemolytic activity (Table 1) [33,36]. However, it is possible that hemolysis inhibition is caused by the irreversible oligomerization of the toxin on sterol aggregates (as shown previously for cholesterol) [33], and not by the competition between the sterol and cholesterol for the binding site. The correlation between hemolysis inhibition and cholesterol binding in aqueous buffer is more apparent than the one observed with liposomes. As mentioned above, the interaction of the sterols with lipids may complicate the interpretation of sterol effect on toxin binding when using lipid bilayers.

Binding of a water-soluble PFO monomer to the membrane is diffusional and electrostatic interactions may play a role since it has been observed that the introduction or elimination of charged residues alters binding [12,16,24,25,49]. While on the membrane surface, insertion of non-polar and aromatic amino acids, and/or specific interactions with membrane lipids, help to anchor the protein to the membrane [50]. However, hydrophobic amino acids are rarely exposed on surfaces of water-soluble proteins, and therefore conformational changes are required to facilitate the interaction of these residues with the hydrophobic core of the membrane bilayer. These conformational changes may contribute to generate a non-polar cavity required to fit a cholesterol molecule. Cholesterol has been found located into non-polar protein cavities, for example for the lysosomal protein NPC2, responsible for Niemann-Pick type C disease [51]. No such cavity is apparent in the structure of the water-soluble monomer of PFO.

The binding model, which was produced via a systematic survey of conformational states of undecapeptide (offered by solved crystal structures of related toxins, Figure 7) and flexible docking, provided a basis for structure-guided rationalization of the cholesterol SAR trends reported herein. Thus, the conserved Thr-Leu pair from L1, essential for recognition of cholesterol, is engaged in the model by the bound ligand, while residues within the undecapeptide, interact with cholesterol via H-bonding contacts through its backbone (Figure 8). The aliphatic–aliphatic contact predicted by the model to be established between cholesterols' tail and conserved residues at the junction of two β-sheets may accounts for double bond intolerance in the tail, which is expected to reduce conformational flexibility and interfere with the compact interdigitation that the saturated variant is capable of.

Both the pH effect in association and the critical role for the terminal Arg in the undecapeptide can be rationalized by the decoupling of the E458/R468 ionic pair (in PFO, see Figure 8) upon cholesterol recognition. This leads to change in glutamate's pKa that promotes protonation and reduction in surface polarity of the cholesterol-binding site. The exposed Arg is, in turn, likely important for electrostatic contact with the anionic head-groups of membrane lipids.

This binding model also provides the basis for some of the cholesterol-dependent effects observed when residues in the D4 loops are mutated. For example, T490 in the PFO-cholesterol model is predicted to be involved in a complex H-bonding network involving both its side chain and backbone carbonyl and the side chains of T460 and E458 (in its protonated form) from the undecapeptide. The static model, however, does not provide a simple explanation for the adjacent L491S mutation that does not significantly change the cholesterol-binding of PFO [25]. This implies a newly found H-bonding role for the Ser side chain at this position that, in contrast with the reduced affinity observed for the L491A substitution, conserved the affinity of PFO for cholesterol.

Finally, the cholesterol-assisted quenching of H-bonding capacity of the flexible undecapeptide may have profound outcome on passive diffusibility of this peptide for anchoring at the membrane. Attainment of conformations that allow quenching of H-bonding capacity by peptide bond NH groups has been noted as a critical pre-requisite for passive internalization of cyclic peptides into biological membranes [52]. The energetic basis for this requirement is found in high energy of desolvation of fully exposed peptide bonds upon passage from high-dielectric water to the low-dielectric interior of a membrane that is typically associated with poor membrane permeation of unstructured peptides in general. Hence, the conformational adaptation of the undecapeptide upon association with cholesterol predicted by the model accomplishes several transformations that combine to promote anchoring at the membrane: (i) higher density of hydrophobic residues projected toward the membrane; (ii) enhanced charge complementarity in the form of exposed R468; and (iii) reduced desolvation costs for the internalization of undecapeptide via H-bonding quench with the cholesterol's hydroxyl. While this model would require further experimental validation, it offers new insights into the D4–cholesterol interaction than can be capitalized in future studies.

4. Materials and Methods

4.1. Materials

Phospholipids were from Avanti Polar Lipids); β-methyl-cyclodextrin (mCD) from Sigma C-4555 (mean degree of substitution 10.5–14.7), 7-dehydrocholesterol higher than 98% by HPLC Fluka, ergosterol 98% by HPLC Fluka, stigmasterol ~95% GC Fluka, β-sitosterol higher than 98%, and cholesterol from Steraloids.

Preparation of nPFO and rPFO$^{V322C-NBD}$ derivatives was done as described previously [33]. rPFO refers to the use of the Cys less PFO^{C459A} derivative.

4.2. Incubation with Sterol Dispersions in Aqueous Solutions

Water-soluble PFO samples (0.3 mL final volume, 0.1 μM final concentration) in buffer C (50 mM Hepes (pH 7.5), 100 mM NaCl, 0.5 mM EDTA, 1 mM DTT) were equilibrated at 37 °C for 5 min before the net initial emission intensity (F_0) was determined (i.e., after blank subtraction). Sterols were then added to the indicated final concentration, and the sample was then incubated at 37 °C for 15 min. The net emission intensity (F) of the sample was determined after blank subtraction and dilution correction. Sterols were dissolved in absolute ethanol to 10 mM and diluted with additional ethanol as necessary. When added to solutions of nPFO or rPFO derivatives, the final concentration of ethanol was always lower than 5% (v/v). Control samples were incubated with an identical volume of ethanol. When indicated, sterol aggregates were dissolved by the addition of mCD to a final concentration of 3 mM. mCD was prepared dissolving 83 mg into 1 mL of HBS buffer (Hepes 50 mM pH 7.5, NaCl 100 mM) at 37 °C for 15 min, centrifuged full speed in microfuge and filtered using a 0.22 μm Millipore filter. The solution was stored at 4 °C for no more than a month.

4.3. Steady-State Fluorescence Spectroscopy

Intensity measurements were performed using the same instrumentation described earlier [25,33,53]. The excitation wavelength and bandpass, and the emission wavelength and bandpass, were, respectively: 470, 4, 530, and 4 nm for NBD; 295, 2, 348, and 4 nm for Trp; 470; and 500, 1, and 500, 1 nm for right angle light scattering measurements.

4.4. Molecular Modeling

All computational procedures were carried out using Schrödinger's Small-Molecule Drug Discovery suite of programs (v. 2016-1, Schrodinger, LLC, New York, NY, USA): Maestro, Protein Preparation Wizard, Epik, Glide, Prime and Induced Fit. The energy optimized all-atom models were generated via a protonation state assignment (Epik), missing atom/loop reconstitution (Prime, OPLS3 force field) and constrained minimization (Prime) sequence within Maestro's Protein Preparation Wizard. The flexible docking was initiated by placing the entire loop-region of D4 into a $15 \times 15 \times 15$ Å^3 grid box. Residues within 5 Å of ligand poses obtained with side-chain-free models (Glide) were refined (Prime) through docking-minimization iterations (Induced Fit).

Supplementary Materials: The following are available online at www.mdpi.com/2072-6651/9/12/381/s1, Figure S1: Absorbance spectra of sterols in ethanol; Figure S2: Sequence and structure similarity between D4 domains of PFO and PLY.

Acknowledgments: This work was supported by Grant Number GM097414 from the National Institute of Health (A.P.H.). Its contents are solely the responsibility of the authors and do not necessarily represent the official views of the National Institute of General Medical Sciences. We thank M. Breña for reading the manuscript and providing valuable feedback.

Author Contributions: A.P.H. conceived and designed the experiments; A.P.H. performed the experiments; S.N.S. performed the bioinformatics and modeling analyses; S.N.S. and A.P.H. analyzed the data; S.N.S and A.P.H. wrote the paper.

Conflicts of Interest: The authors declare no conflict of interest. The founding sponsors had no role in the design of the study; in the collection, analyses, or interpretation of data; in the writing of the manuscript, and in the decision to publish the results.

References

1. Tweten, R.K. Cholesterol-dependent cytolysins, a family of versatile pore-forming toxins. *Infect. Immun.* **2005**, *73*, 6199–6209. [CrossRef] [PubMed]
2. Heuck, A.P.; Moe, P.C.; Johnson, B.B. The cholesterol-dependent cytolysins family of Gram-positive bacterial toxins. In *Cholesterol Binding Proteins and Cholesterol Transport*; Harris, J.R., Ed.; Springer: Dordrecht, The Netherlands, 2010; Volume 51, pp. 551–577.

3. Johnson, B.; Heuck, A. Perfringolysin O structure and mechanism of pore formation as a paradigm for cholesterol-dependent cytolysins. In *Macpf/cdc Proteins—Agents of Defence, Attack and Invasion*; Anderluh, G., Gilbert, R., Eds.; Springer: Dordrecht, The Netherlands, 2014; Volume 80, pp. 63–81.

4. Giddings, K.S.; Zhao, J.; Sims, P.J.; Tweten, R.K. Human CD59 is a receptor for the cholesterol-dependent cytolysin intermedilysin. *Nat. Struct. Mol. Biol.* **2004**, *11*, 1173–1178. [CrossRef] [PubMed]

5. Giddings, K.S.; Johnson, A.E.; Tweten, R.K. Redefining cholesterol's role in the mechanism of the cholesterol-dependent cytolysins. *Proc. Natl. Acad. Sci. USA* **2003**, *100*, 11315–11320. [CrossRef] [PubMed]

6. Alving, C.R.; Habig, W.H.; Urban, K.A.; Hardegree, M.C. Cholesterol-dependent tetanolysin damage to liposomes. *Biochim. Biophys. Acta* **1979**, *551*, 224–228. [CrossRef]

7. Rosenqvist, E.; Michaelsen, T.E.; Vistnes, A.I. Effect of streptolysin O and digitonin on egg lecithin/cholesterol vesicles. *Biochim. Biophys. Acta* **1980**, *600*, 91–102. [CrossRef]

8. Ohno-Iwashita, Y.; Iwamoto, M.; Ando, S.; Iwashita, S. Effect of lipidic factors on membrane cholesterol topology—Mode of binding of θ-toxin to cholesterol in liposomes. *Biochim. Biophys. Acta* **1992**, *1109*, 81–90. [CrossRef]

9. Nelson, L.D.; Johnson, A.E.; London, E. How interaction of perfringolysin O with membranes is controlled by sterol structure, lipid structure, and physiological low ph: Insights into the origin of perfringolysin O-lipid raft interaction. *J. Biol. Chem.* **2008**, *283*, 4632–4642. [CrossRef] [PubMed]

10. Flanagan, J.J.; Tweten, R.K.; Johnson, A.E.; Heuck, A.P. Cholesterol exposure at the membrane surface is necessary and sufficient to trigger perfringolysin O binding. *Biochemistry* **2009**, *48*, 3977–3987. [CrossRef] [PubMed]

11. Ramachandran, R.; Heuck, A.P.; Tweten, R.K.; Johnson, A.E. Structural insights into the membrane-anchoring mechanism of a cholesterol-dependent cytolysin. *Nat. Struct. Mol. Biol.* **2002**, *9*, 823–827. [CrossRef] [PubMed]

12. Soltani, C.E.; Hotze, E.M.; Johnson, A.E.; Tweten, R.K. Structural elements of the cholesterol-dependent cytolysins that are responsible for their cholesterol-sensitive membrane interactions. *Proc. Natl. Acad. Sci. USA* **2007**, *104*, 20226–20231. [CrossRef] [PubMed]

13. Farrand, A.J.; LaChapelle, S.; Hotze, E.M.; Johnson, A.E.; Tweten, R.K. Only two amino acids are essential for cytolytic toxin recognition of cholesterol at the membrane surface. *Proc. Natl. Acad. Sci. USA* **2010**, *107*, 4341–4346. [CrossRef] [PubMed]

14. Polekhina, G.; Giddings, K.S.; Tweten, R.K.; Parker, M.W. Insights into the action of the superfamily of cholesterol-dependent cytolysins from studies of intermedilysin. *Proc. Natl. Acad. Sci. USA* **2005**, *102*, 600–605. [CrossRef] [PubMed]

15. Moe, P.C.; Heuck, A.P. Phospholipid hydrolysis caused by *Clostridium perfringens* α-toxin facilitates the targeting of perfringolysin o to membrane bilayers. *Biochemistry* **2010**, *49*, 9498–9507. [CrossRef] [PubMed]

16. Johnson, B.B.; Moe, P.C.; Wang, D.; Rossi, K.; Trigatti, B.L.; Heuck, A.P. Modifications in perfringolysin O domain 4 alter the cholesterol concentration threshold required for binding. *Biochemistry* **2012**, *51*, 3373–3382. [CrossRef] [PubMed]

17. Lally, C.C.M.; Bauer, B.; Selent, J.; Sommer, M.E. C-edge loops of arrestin function as a membrane anchor. *Nat. Commun.* **2017**, *8*, 14258. [CrossRef] [PubMed]

18. Saunders, F.K.; Mitchell, T.J.; Walker, J.A.; Andrew, P.W.; Boulnois, G.J. Pneumolysin, the thiol-activated toxin of *Streptococcus pneumoniae*, does not require a thiol group for in vitro activity. *Infect. Immun.* **1989**, *57*, 2547–2552. [PubMed]

19. Pinkney, M.; Beachey, E.; Kehoe, M. The thiol-activated toxin streptolysin O does not require a thiol group for cytolytic activity. *Infect. Immun.* **1989**, *57*, 2553–2558. [PubMed]

20. Michel, E.; Reich, K.A.; Favier, R.; Berche, P.; Cossart, P. Attenuated mutants of the intracellular bacterium *Listeria monocytogenes* obtained by single amino acid substitutions in listeriolysin O. *Mol. Microbiol.* **1990**, *4*, 2167–2178. [CrossRef] [PubMed]

21. Sekino-Suzuki, N.; Nakamura, M.; Mitsui, K.-I.; Ohno-Iwashita, Y. Contribution of individual tryptophan residues to the structure and activity of θ-toxin (perfringolysin O), a cholesterol-binding cytolysin. *Eur. J. Biochem.* **1996**, *241*, 941–947. [CrossRef] [PubMed]

22. Korchev, Y.E.; Bashford, C.L.; Pederzolli, C.; Pasternak, C.A.; Morgan, P.J.; Andrew, P.W.; Mitchell, T.J. A conserved tryptophan in pneumolysin is a determinant of the characteristics of channels formed pneumolysin in cells and planar lipid bilayers. *Biochem. J.* **1998**, *329*, 571–577. [CrossRef] [PubMed]

23. Billington, S.J.; Songer, J.G.; Jost, B.H. The variant undecapeptide sequence of the *Arcanobacterium pyogenes* haemolysin, pyolysin, is required for full cytolytic activity. *Microbiology* **2002**, *148*, 3947–3954. [CrossRef] [PubMed]

24. Dowd, K.J.; Tweten, R.K. The cholesterol-dependent cytolysin signature motif: A critical element in the allosteric pathway that couples membrane binding to pore assembly. *PLoS Pathog.* **2012**, *8*, e1002787. [CrossRef]

25. Johnson, B.B.; Breña, M.; Anguita, J.; Heuck, A.P. Mechanistic insights into the cholesterol-dependent binding of perfringolysin O-based probes and cell membranes. *Sci. Rep.* **2017**, *7*, 13793. [CrossRef] [PubMed]

26. Park, S.A.; Park, Y.S.; Bong, S.M.; Lee, K.S. Structure-based functional studies for the cellular recognition and cytolytic mechanism of pneumolysin from *Streptococcus pneumoniae*. *J. Struct. Biol.* **2016**, *193*, 132–140. [CrossRef] [PubMed]

27. Heuck, A.P.; Hotze, E.M.; Tweten, R.K.; Johnson, A.E. Mechanism of membrane insertion of a multimeric β-barrel protein: Perfringolysin O creates a pore using ordered and coupled conformational changes. *Mol. Cell* **2000**, *6*, 1233–1242. [CrossRef]

28. Bavdek, A.; Gekara, N.O.; Priselac, D.; Gutierrez Aguirre, I.; Darji, A.; Chakraborty, T.; Macĺœek, P.; Lakey, J.H.; Weiss, S.; Anderluh, G. Sterol and pH interdependence in the binding, oligomerization, and pore formation of listeriolysin O. *Biochemistry* **2007**, *46*, 4425–4437. [CrossRef] [PubMed]

29. Waheed, A.; Shimada, Y.; Heijnen, H.F.G.; Nakamura, M.; Inomata, M.; Hayashi, M.; Iwashita, S.; Slot, J.W.; Ohno-Iwashita, Y. Selective binding of perfringolysin O derivative to cholesterol-rich membrane microdomains (rafts). *Proc. Natl. Acad. Sci. USA* **2001**, *98*, 4926–4931. [CrossRef] [PubMed]

30. Ohno-Iwashita, Y.; Shimada, Y.; Waheed, A.; Hayashi, M.; Inomata, M.; Nakamura, M.; Maruya, M.; Iwashita, M. Perfringolysin O, a cholesterol-binding cytolysin, as a probe for lipid rafts. *Anaerobe* **2004**, *10*, 125–134. [CrossRef] [PubMed]

31. Sokolov, A.; Radhakrishnan, A. Accessibility of cholesterol in endoplasmic reticulum membranes and activation of SREBP-2 switch abruptly at a common cholesterol threshold. *J. Biol. Chem.* **2010**, *285*, 29480–29490. [CrossRef] [PubMed]

32. Olsen, B.N.; Bielska, A.A.; Lee, T.; Daily, M.D.; Covey, D.F.; Schlesinger, P.H.; Baker, N.A.; Ory, D.S. The structural basis of cholesterol accessibility in membranes. *Biophys. J.* **2013**, *105*, 1838–1847. [CrossRef] [PubMed]

33. Heuck, A.P.; Savva, C.G.; Holzenburg, A.; Johnson, A.E. Conformational changes that effect oligomerization and initiate pore formation are triggered throughout perfringolysin O upon binding to cholesterol. *J. Biol. Chem.* **2007**, *282*, 22629–22637. [CrossRef] [PubMed]

34. Prigent, D.; Alouf, J.E. Interaction of streptolysin o with sterols. *Biochim. Biophys. Acta* **1976**, *443*, 288–300. [CrossRef]

35. Kenneth, C.; Watson, K.C.; Kerr, E.J. Sterol structural requirements for inhibition of streptolysin O activity. *Biochem. J.* **1974**, *140*, 95–98.

36. Hase, J.; Mitsui, K.; Shonaka, E. *Clostridium perfringens* exotoxins. Iv. Inhibition of the theta-toxin induced hemolysis by steroids and related compounds. *Jpn. J. Exp. Med.* **1976**, *46*, 45–50. [PubMed]

37. Alouf, J.E. *Streptococcal* toxins (streptolysin O, streptolysin S, erythrogenic toxin). *Pharmacol. Ther.* **1980**, *11*, 661–717. [CrossRef]

38. Megha; Bakht, O.; London, E. Cholesterol precursors stabilize ordinary and ceramide-rich ordered lipid domains (lipid rafts) to different degrees: Implications for the bloch hypothesis and sterol biosynthesis disorders. *J. Biol. Chem.* **2006**, *281*, 21903–21913.

39. Haberland, M.E.; Reynolds, J.A. Self-association of cholesterol in aqueous solution. *Proc. Natl. Acad. Sci. USA* **1973**, *70*, 2313–2316. [CrossRef] [PubMed]

40. Gimpl, G.; Burger, K.; Fahrenholz, F. Cholesterol as modulator of receptor function. *Biochemistry* **1997**, *36*, 10959–10974. [CrossRef] [PubMed]

41. Nelson, L.D.; Chiantia, S.; London, E. Perfringolysin O association with ordered lipid domains: Implications for transmembrane protein raft affinity. *Biophys. J.* **2010**, *99*, 3255–3263. [CrossRef] [PubMed]

42. Ramachandran, R.; Tweten, R.K.; Johnson, A.E. Membrane-dependent conformational changes initiate cholesterol-dependent cytolysin oligomerization and intersubunit beta-strand alignment. *Nat. Struct. Mol. Biol.* **2004**, *11*, 697–705. [CrossRef] [PubMed]

43. Sherman, W.; Day, T.; Jacobson, M.P.; Friesner, R.A.; Farid, R. Novel procedure for modeling ligand/receptor induced fit effects. *J. Med. Chem.* **2006**, *49*, 534–553. [CrossRef] [PubMed]

44. Feil, S.C.; Ascher, D.B.; Kuiper, M.J.; Tweten, R.K.; Parker, M.W. Structural studies of *Streptococcus pyogenes* streptolysin O provide insights into the early steps of membrane penetration. *J. Mol. Biol.* **2014**, *426*, 785–792. [CrossRef] [PubMed]

45. Gilbert, R. Structural features of cholesterol dependent cytolysins and comparison to other MACPF-domain containing proteins. In *MACPF/CDC Proteins—Agents of Defence, Attack and Invasion*; Anderluh, G., Gilbert, R., Eds.; Springer: Dordrecht, The Netherlands, 2014; Volume 80, pp. 47–62.

46. Chou, P.Y.; Fasman, G.D. Empirical predictions of protein conformation. *Ann. Rev. Biochem.* **1978**, *47*, 251–276. [CrossRef] [PubMed]

47. Kulma, M.; Kacprzyk-Stokowiec, A.; Kwiatkowska, K.; Traczyk, G.; Sobota, A.; Dadlez, M. R468A mutation in perfringolysin O destabilizes toxin structure and induces membrane fusion. *Biochim. Biophys. Acta (BBA) Biomembr.* **2017**, *1859*, 1075–1088. [CrossRef] [PubMed]

48. Shelley, J.C.; Cholleti, A.; Frye, L.L.; Greenwood, J.R.; Timlin, M.R.; Uchimaya, M. Epik: A software program for pKa prediction and protonation state generation for drug-like molecules. *J. Comput.-Aided Mol. Des.* **2007**, *21*, 681–691. [CrossRef] [PubMed]

49. Farrand, A.J.; Hotze, E.M.; Sato, T.K.; Wade, K.R.; Wimley, W.C.; Johnson, A.E.; Tweten, R.K. The cholesterol-dependent cytolysin membrane-binding interface discriminates lipid environments of cholesterol to support β-barrel pore insertion. *J. Biol. Chem.* **2015**, *290*, 17733–17744. [CrossRef] [PubMed]

50. Cho, W.; Stahelin, R.V. Membrane-protein interactions in cell signaling and membrane trafficking. *Annu. Rev. Biophys. Biomol. Struct.* **2005**, *34*, 119–151. [CrossRef] [PubMed]

51. Xu, S.; Benoff, B.; Liou, H.-L.; Lobel, P.; Stock, A.M. Structural basis of sterol binding by NPC2, a lysosomal protein deficient in Niemann-Pick typeC2 disease. *J. Biol. Chem.* **2007**, *282*, 23525–23531. [CrossRef] [PubMed]

52. Rezai, T.; Bock, J.E.; Zhou, M.V.; Kalyanaraman, C.; Lokey, R.S.; Jacobson, M.P. Conformational flexibility, internal hydrogen bonding, and passive membrane permeability: Successful in silico prediction of the relative permeabilities of cyclic peptides. *J. Am. Chem. Soc.* **2006**, *128*, 14073–14080. [CrossRef] [PubMed]

53. Shepard, L.A.; Heuck, A.P.; Hamman, B.D.; Rossjohn, J.; Parker, M.W.; Ryan, K.R.; Johnson, A.E.; Tweten, R.K. Identification of a membrane-spanning domain of the thiol-activated pore-forming toxin *Clostridium perfringens* perfringolysin O: An α-helical to β-sheet transition identified by fluorescence spectroscopy. *Biochemistry* **1998**, *37*, 14563–14574. [CrossRef] [PubMed]

toxins

MDPI

Article

Asymmetric Cryo-EM Structure of Anthrax Toxin Protective Antigen Pore with Lethal Factor N-Terminal Domain

Alexandra J. Machen [1,†], Narahari Akkaladevi [2,†], Caleb Trecazzi [1], Pierce T. O'Neil [1], Srayanta Mukherjee [1], Yifei Qi [3], Rebecca Dillard [4], Wonpil Im [3], Edward P. Gogol [5], Tommi A. White [2,6] and Mark T. Fisher [1,*]

[1]   Department of Biochemistry and Molecular Biology, University of Kansas Medical Center, Kansas City, KS 66160, USA; amachen@kumc.edu (A.J.M.); ctrecazzi@kumc.edu (C.T.); poneil@kumc.edu (P.T.O.); srayanta@gmail.com (S.M.)
[2]   Department of Biochemistry, University of Missouri, Columbia, MO 65211, USA; akkaladevin@missouri.edu (N.A.); whiteto@missouri.edu (T.A.W.)
[3]   Departments of Biological Sciences and Bioengineering, Lehigh University, Bethlehem, PA 18015, USA; yfqi@chem.ecnu.edu.cn (Y.Q.); woi216@lehigh.edu (W.I.)
[4]   National Center for Macromolecular Imaging, Baylor University, Houston TX 77030, USA; rdillard@gmail.com
[5]   Department of Biological Sciences, University of Missouri-Kansas City, Kansas City, MO 64110, USA; gogole@umkc.edu
[6]   Electron Microscopy Core Facility, University of Missouri, Columbia, MO 65211, USA
*   Correspondence: mfisher1@kumc.edu; Tel.: +1-913-588-6940
†   These authors contributed to this work equally.

Academic Editor: Michel R. Popoff
Received: 17 August 2017; Accepted: 19 September 2017; Published: 22 September 2017

Abstract: The anthrax lethal toxin consists of protective antigen (PA) and lethal factor (LF). Understanding both the PA pore formation and LF translocation through the PA pore is crucial to mitigating and perhaps preventing anthrax disease. To better understand the interactions of the LF-PA engagement complex, the structure of the LF_N-bound PA pore solubilized by a lipid nanodisc was examined using cryo-EM. CryoSPARC was used to rapidly sort particle populations of a heterogeneous sample preparation without imposing symmetry, resulting in a refined 17 Å PA pore structure with 3 LF_N bound. At pH 7.5, the contributions from the three unstructured LF_N lysine-rich tail regions do not occlude the Phe clamp opening. The open Phe clamp suggests that, in this translocation-compromised pH environment, the lysine-rich tails remain flexible and do not interact with the pore lumen region.

Keywords: anthrax toxin; lethal factor; protective antigen; pore formation; translocation; nanodisc; cryo-EM; cryoSPARC

1. Introduction

The lethality of anthrax, a zoonotic disease and bioterrorism agent, is due to the anthrax toxin. This tripartite toxin consists of a protective antigen (PA), lethal factor (LF; a mitogen-activated protein kinase kinase protease), and edema factor (EF; an adenylate cyclase) [1]. After secretion from *Bacillus anthracis*, the 83 kDa PA (PA_{83}) binds to its target host cell receptor, either capillary morphogenesis protein 2 (CMG2) or tumor endothelium marker-8 (TEM8) [1–4]. PA_{83} is cleaved by proteases, resulting in 20 kDa and 63 kDa fragments. PA_{63} then self-associates to form a heptameric PA prepore that can associate with up to three molecules of LF or EF [5]. Octameric PA prepores may also assemble in solution, governed by LF or EF binding to PA_{63} monomers clipped in solution [6]. Receptor-bound assembled complexes are endocytosed. As the endosome acidifies to pH 5.0 (late endosome), the receptor-bound PA prepore undergoes a conformational change into an extended β-barrel pore structure that penetrates the endosomal membrane. This newly-formed structure facilitates unfolding and translocation of the 90 kDa LF (or EF) enzyme across the pH gradient of the endosomal membrane through the narrow PA pore lumen in a pH-driven hypothesized Brownian ratchet mechanism [7]. Translocation of α-helical regions of LF are aided by the PA α-clamp [8,9]. LF translocation is gated by a ring of seven phenylalanine residues, termed the Phe clamp, located further down the PA pore lumen [10–13]. The directional translocation of LF depends on protonation of acidic residues, the electrostatic character of the PA pore lumen, and any residual positive charges on LF [14]. Subsequent deprotonation of the translocating peptide after passing through the Phe clamp prevents back transfer. Translocated LF refolds on the cytosolic side of the endosomal membrane, where it disrupts cell signaling by cleaving MAP kinase kinases, resulting in cell death [15].

Previously, the Krantz group published work on the large-scale rearrangement of LF that occurs upon binding to the PA prepore. Specifically, the N-terminal α-helix of LF moves away from the main body of LF and is resituated into a groove in the interior surface of the PA prepore cap, termed the α-clamp region [16]. This reposition is proposed to help funnel the N-termini of LF into the PA pore lumen. The narrowest part of the pore lumen is the Phe clamp [10,17]. PA F427, which forms the heptameric Phe clamp, is an essential residue that facilitates LF translocation [12,18]. Mutations in this residue (e.g., F427A) affect the kinetics of pore formation and translocation [19]. Interestingly, the loop containing F427 ($2\beta_{10}$–$2\beta_{11}$ loop) was suggested by Jiang et al. [17] to be involved in the first unfolding step of the pore formation mechanism. This mechanism is based on a comparisons of crystal structures of the oligomeric PA prepore [20] and the 2.9 Å cryo-EM pore structure [17]. The $2\beta_{10}$–$2\beta_{11}$ loop also contains D426 which forms a conserved inter-subunit salt bridge with K397 in the PA pore [11]. These interactions orient F427 into its constricted Phe clamp formation, which is hypothesized as pi-stacking interactions between adjacent F427 residues [1,11]. The first step of this pore forming mechanism is based on the increased flexibility of the $2\beta_{10}$–$2\beta_{11}$ loop in various prepore crystal structures. Early characterization of LF-PA interactions showed the N-terminal tail of LF interacts with the Phe clamp of the PA pore at pH 5.0, which has since been verified using cysteine cross-linking [12,13].

PA pore formation is necessary, but not sufficient, for lethality: LF must be translocated through the pore. Das and Krantz [9] recently proposed the Phe clamp region is dynamic and can undergo large-scale movements to momentarily increase the pore diameter from 6 Å to 10–12 Å. These movements could resemble transient open forms due to salt bridge formation between the acidic residues in the Phe clamp loop and an adjacent monomer [11]. This latter conformation (open Phe clamp loop) affects translocation rates since mutations that inhibit salt bridge formation impact translocation kinetics. In this particular model, Krantz also presented single-channel evidence that α-helical structures translocate more efficiently than extended β-sheet-like structures or unstructured polypeptides containing alternating L- and D-amino acids. Pentelute et al. [21] showed chirality is not important for translocation of the unstructured region of the N-terminal domain of LF (LF$_N$) either, but this does not preclude the possibility that α-helical structures could be formed upon the electrostatic interaction between LF$_N$ and the PA pore. This would have to include α-helices that are in the D chiral form since all L- or D-amino acid α-helical structures do not slow down translocation. It would then be of value to determine if the Phe clamp loop region becomes more structurally dynamic (loss of resolution) and/or adapts a more open configuration upon interaction with the single or multiple unstructured lysine-rich tails of bound lethal factor(s).

Understanding both pore formation and LF translocation is imperative in order to develop strategies that mitigate or prevent the formation of the anthrax toxin complex or inhibit the translocation mechanism. Inhibition of circulating anthrax toxins is crucial since the toxin components retain cell lethality even after the bacilli have been killed with antibiotics [1]. To better understand the interactions between LF$_N$ and PA, the structure of the LF$_N$-bound PA pore in a lipid membrane environment was examined using cryo-electron microscopy (cryo-EM).

2. Results

2.1. Cryo-EM Sample Preparation of PA Pore with Three LF$_N$ Bound

With the recent publication of the cryo-EM PA pore structure at pH 5.0 [17], the logical, but challenging, next step in understanding anthrax toxin pore formation and translocation involves determining how bound LF influences the conformation of the PA pore. An atomic resolution structure of LF$_N$-bound PA pore would give molecular insight into the nuances of this interaction. In order to solve the cryo-EM structure of the LF$_N$-PA pore, several obstacles must be overcome, including the aggregation propensity of the pore, asymmetry of the LF$_N$-PA complex, and orientational preferences of complexes on EM grids.

We previously published a methodology to assemble LF$_N$-PA pore complexes while avoiding aggregation by immobilizing PA pores before solubilizing the hydrophobic tip with lipid bilayer nanodiscs [22–25]. After immobilization, the PA prepores were transitioned into pores using a urea/37°C pulse methodology, exposing the aggregation-prone pore tip. The nanodisc formed around the hydrophobic pore tip while the complex was immobilized [22,23,25,26]. A schematic of this methodology is shown in Figure 1A. Our previous low-resolution LF$_N$-PA-nanodisc structures were reconstructed from samples frozen on perforated carbon containing a thin carbon layer over holes [23]. There were a number of caveats limiting the structural analysis of that preparation. Most importantly, large diameter nanodiscs (approximately 400 Å) were generated and required the use of thicker ice. In addition, LF$_N$-PA-nanodisc complexes interacted with the carbon layer, resulting in complexes preferring a side view orientation which displays the long axis of the heptameric PA pore rather than allowing for more diverse conformational orientations, including top views. Although these LF$_N$-PA-nanodisc complexes were inherently structurally asymmetric (symmetry mismatch of seven PA subunits to a maximum of three LF$_N$ bound), their structures were generated by imposing seven-fold symmetry, which resulted in smearing of the LF$_N$-bound density. This coupled with the sample-induced constraints (thin carbon backing, thick ice, and Fresnel fringe effects for the sharp nanodisc protein interface) diminished the contrast of the protein. These constraints also interfered with the visualization of the PA β-barrel in the reconstruction.

To obtain a more concise LF$_N$-PA-nanodisc complex structure, these sample preparation issues had to be overcome. For better contrast, samples were frozen on simple perforated carbon grids without a thin carbon layer in order to achieve greater orientational diversity and were imaged with a JEM 2200FS electron microscope (60,000× magnification) (JEOL, Peabody, MA, USA). A representative micrograph with high defocus for better contrast (for visualization, not reconstruction purposes) with individual particles highlighted with red circles is shown in Figure 1B. Low-dose, low-defocus conditions were used to collect images for 3D reconstruction. Notably, the nanodiscs for these samples were significantly smaller than the previous larger nanodisc samples. The nanodisc size was dependent on the length of time that LF$_N$-PA-nanodisc complexes were immobilized as well as rotation of the sample tube. Under non-ideal conditions, the pre-nanodisc micelles may merge, generating larger nanodisc diameters. Interestingly, larger nanodiscs often resulted in multiple PA pore-inserted nanodisc complexes (e.g., sometimes four PA pores inserted into one large nanodisc). These larger nanodiscs were attributed to longer dialysis times that consistently resulted in merging of pre-nanodisc micelles. Reducing the time of incubation, ensuring adequate detergent dialysis with Bio-Beads, and constant rotation during formation yielded smaller nanodiscs within the expected diameter range (100–150 Å) containing a single PA pore (Figure 1B).

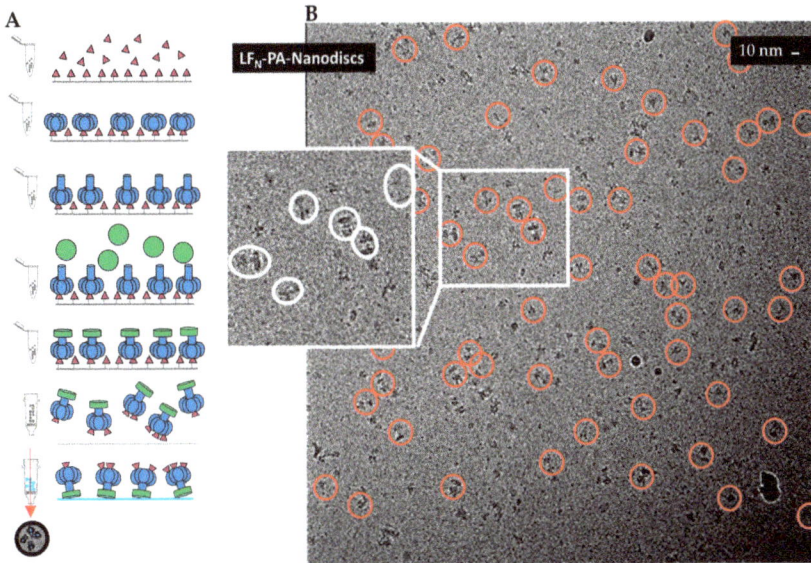

Figure 1. Sample preparation of 3LF$_N$-PA-nanodiscs: (**A**) schematic of LF$_N$ (magenta)-PA (blue)-nanodisc (green) complex formation with stepwise addition of LF$_N$ and PA to thiol sepharose beads; and (**B**) a higher defocus representative field for high-contrast visualization. Individual LF$_N$-PA-nanodisc complexes may be easily observed within this micrograph. Note the variable size of the nanodiscs in the insert.

2.2. Single-Particle Analysis of LF$_N$-PA-Nanodisc Complexes

Initial classification analysis using SPARX [27] revealed heterogeneity in the dataset with one, two, or three LF$_N$ bound to PA pores (Figure 2A–C). The release of LF$_N$-PA-nanodisc complexes from the bead surface into solution also resulted in the release of non-complexed LF$_N$, which was then able to bind released complexes leading to particles with multiple binding events. This led to subsets of PA having one, two, or three LF$_N$ bound. This inherent heterogeneity in LF$_N$ binding stoichiometry made 3D reconstruction difficult. Initially, this limited particle dataset could only be used to obtain a model by imposing C7 symmetry during reconstruction using EMAN2.1 and RELION (Figure 2D).

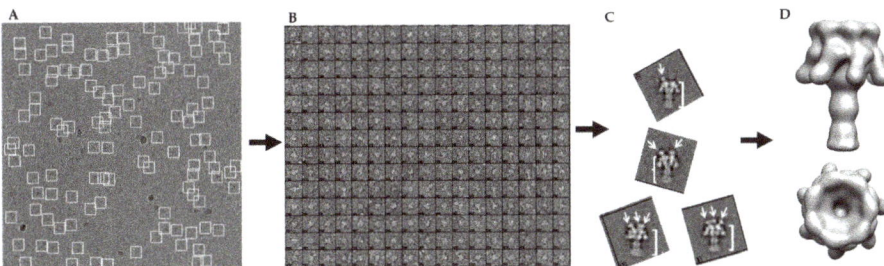

Figure 2. Schematic of data analysis of heterogeneous cryo-EM data: (**A**) sample micrograph field showing good ice and particle distribution; (**B**) an example of individual boxed particles from micrographs with phase inversion for contrast; (**C**) SPARX 2D class averages (side views) reveal heterogeneity of sample preparation with arrows indicating LF$_N$ binding; and (**D**) 3D model of 3LF$_N$-PA pore with C7 symmetry imposed smears LF$_N$ density into a crown around the top of the pore.

While PA alone has C7 symmetry, LF_N-bound PA in a saturated (three LF_N bound) or sub-saturated binding ratio only possesses C1 symmetry. The recent successful high-resolution reconstruction of the PA pore at pH 5.0 by Jiang et al. [17] was accomplished using, primarily, top and side view orientations that were generated by taking advantage of a grid adherence platform. In that sample preparation, the prepore adhered to the carbon layer through its receptor binding interface and the pore transition was accomplished by adjusting the pH of the solution to pH 5.0. Since the pore itself has an axis of seven-fold symmetry, the variable positioning of the side views of the PA pore on the carbon layer were sufficient to cover most of the conformational space to obtain the first high-resolution structure (2.9 Å) of the anthrax toxin pore translocon [17]. With LF_N-PA-nanodisc complexes, the nanodisc insertion procedure permits conformational diversity, which is critical for obtaining a structure without imposing sevenfold symmetry. A direction distribution map, analogous to an Euler angle map, confirmed the orientation of the LF_N-PA-nanodisc particles was conformationally diverse (Figure 3).

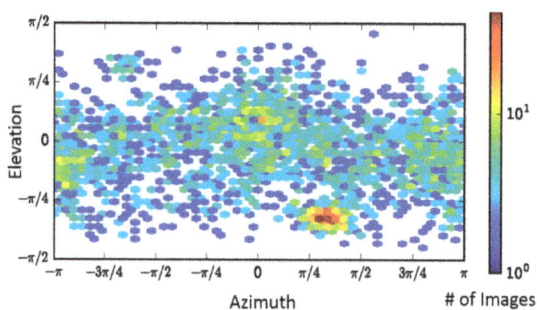

Figure 3. Direction distribution map of particles, analogous to an Euler angle map, showing the conformational coverage of LF_N-PA-nanodiscs.

It is important to note this diverse distribution is crucial for acquiring the asymmetric LF_N-PA-nanodisc structures since the imposition of sevenfold symmetry during 3D reconstructions distorts the density of any bound LF_N (Figure 2D). CryoSPARC is well suited to obtain unbiased, reproducible, and reliable *ab initio* 3D models rapidly even when extensive sample heterogeneity is present [28,29]. For example, Ripstien et al. [30] reexamined their previous cryo-EM data of the *Thermus thermophiles* V/A-ATPase using cryoSPARC and were able to determine their ATPase sample was actually populated by multiple conformations that were previously unresolved, resulting in new mechanistic insights.

To separate the heterogeneous LF_N-PA-nanodisc particles, an initial 2D classification was performed on the 30,696 particles with removal of bad classes as determined by eye (Figure 4A). An *ab initio* classification with four groups was then performed on the remaining 18,806 good particles (Figure 4B). Four groups were chosen since two LF_N can bind to PA at neighboring binding sites or with an empty binding site between them resulting in $1LF_N$, 2_ALF_N, 2_BLF_N, or $3LF_N$ bound. Group 2 was the most highly populated group identified by the cryoSPARC stochastic gradient descent (SGD) *ab initio* model generation with three distinct and equal LF_N densities (Figure 4B). Further 2D classification was performed on all four groups to assess the quality of particles within each group (Figure 4C). Group 3 contained several highly-populated classes showing sharp sevenfold symmetric top and bottom views. Group 1 and 4 particles did not result in clear classes and were discarded (Figure 4C, top and bottom panels). Since the top and bottom view classes in Group 2 were underrepresented, all particles from Group 2 (4560) and particles from the good classes in Group 3 (1159) were combined. A homogeneous refinement was run with the Group 2 *ab initio* model with the combined good particle set (Figure 4D).

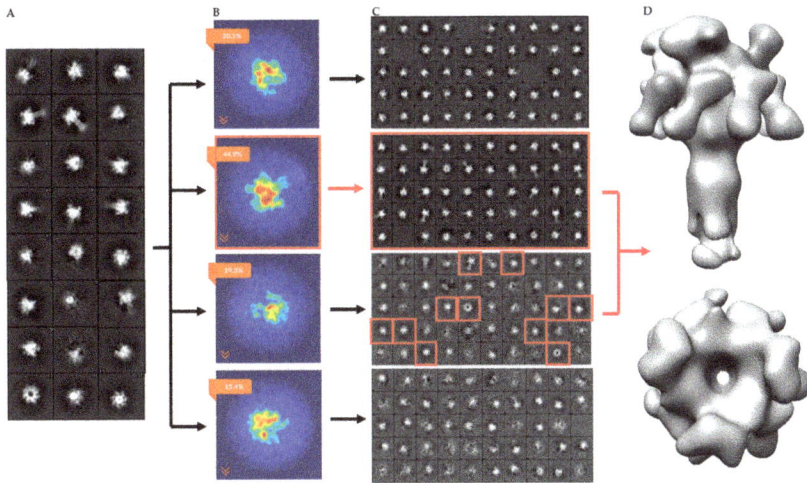

Figure 4. CryoSPARC data analysis flowchart of heterogeneous LF$_N$-PA-nanodiscs with total computational time of 3.5 h from 2D averaging to refined model: (**A**) cryoSPARC 2D class averaging of 18,806 particles; (**B**) image projection of heterogeneous *ab initio* reconstruction with four groups, the largest group, with 44.9% of the particles, corresponds to 3LF$_N$; (**C**) 2D class averages of each *ab initio* particle group; and (**D**) 17 Å model of 3LF$_N$-PA generated from homogeneous refinement of the Group 2 *ab initio* model with particles from top and bottom 2D class averages highlighted in red.

The homogeneous refinement resulted in a 17 Å 3LF$_N$-PA pore model from 5719 particles. Figure 5 shows the Fourier Shell Coefficient (FSC) used to calculate the resolution. This resulting reconstruction was not biased by outside models or symmetrization operations. The β-barrel pore of PA was not prominent in the *ab initio* model but became more apparent upon cryoSPARC refinement. The bulge in the β-barrel of the final model was also seen in the cryo-EM structure of the PA pore alone where this hydrophobic region of the outer barrel bound lipids, resulting in the accumulation of additional density [17]. As can be seen in the 2D classification (Figure 4C, second panel), side view images reveal variation either in nanodisc size or electron density. This resulted in a lack of nanodisc structure in the final electron density map. The irregular density at the bottom of the pore tip in the final structure can be attributed to either the presence of nanodisc or free lipid binding to exposed hydrophobic residues. As mentioned previously, the decrease in nanodisc density appears to be due to extended dialysis times during micelle to nanodisc collapse. The decreased nanodisc size did not diminish our ability to reconstruct LF$_N$-PA pore complexes, particularly in the PA pore cap and the initial extension of the β-barrel.

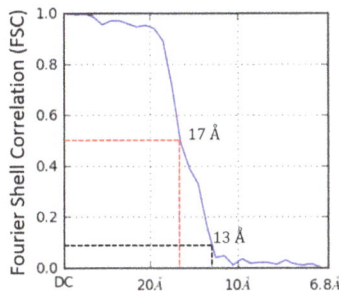

Figure 5. Fourier shell correlation (FSC) for 3LF$_N$-PA-nanodiscs. Resolution estimated to be 17 Å based on FSC with a cutoff of 0.5. This conservative cutoff agrees with filtered models shown later in Figure 10.

2.3. Constructing Samples with Highly-Populated Singly-Bound LF$_N$-PA for Cryo-EM

The heterogeneity of this sample preparation was due to the stepwise assembly of LF$_N$-PA complexes, shown above in Figure 1A. LF$_N$ was immobilized onto thiol sepharose beads, then PA prepore was added, binding to the LF$_N$. The bulkiness of PA relative to LF$_N$ blocked PA from binding to multiple LF$_N$. After LF$_N$-PA-nanodisc complexes were formed on the beads, they were released into solution. Any unbound LF$_N$ was also released and, due to its high affinity for PA, bound to open binding sites of PA (Figure 1A). To obtain a larger, more homogeneous LF$_N$-bound PA pore particle set, the protocol was modified by pre-incubating LF$_N$ with PA prepore in a 1:2 ratio to ensure a higher population of singly-bound LF$_N$-PA. A schematic of the updated protocol is shown in Figure 6A. As proof of principle for future structure determinations, an initial cryo-EM screen of complexes isolated with this new protocol was performed. Figure 6B shows a representative screening image collected on F30 twin TEM (FEI, Hillsboro, OR, USA) at 39,000 times nominal magnification and a pixel size of 3 Å on the specimen. 2D class averaging with SPARX (side views shown in Figure 7) showed the majority of the classified populations had single LF$_N$ densities. As with all preparations using the immobilized construction of LF$_N$-PA pore complexes, the elution volume is easily adjusted to obtain a sufficient concentration of particles on the grid for automated screening with a high-powered microscope with a direct electron detector.

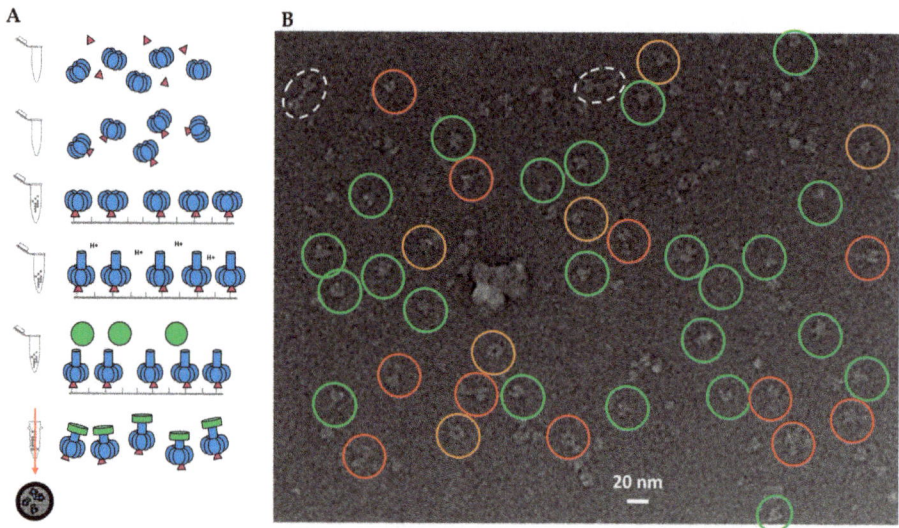

Figure 6. Sample preparation of 1LF$_N$-PA-nanodiscs: (**A**) schematic of LF$_N$ (magenta)-PA (blue)-nanodisc (green) complex formation with LF$_N$ and PA prepore incubated prior to immobilization; and (**B**) representative cryo-EM image field of initial screening. Inverted contrast for visualization. Only select clear individual particles are noted (key: red—side views; green—various angle views; dotted white—double PA pores in a single nanodisc; orange—top and bottom views). Extra density from LF$_N$ binding is occasionally observed, particularly in the side view orientations.

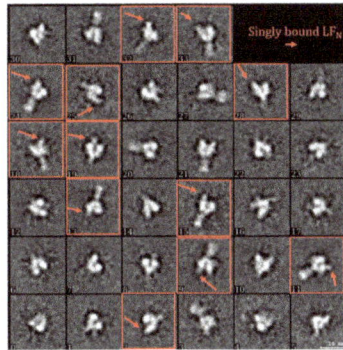

Figure 7. 2D classification of approximately 1200 particles using SPARX confirmed singly-bound LF$_N$ with examples of clear LF$_N$ densities highlighted in red.

2.4. Molecular Dynamics Flexible Fitting of 3LF$_N$-PA Pore Model into the 17 Å Cryo-EM Density Map

The refined 17 Å cryo-EM model of 3LF$_N$-PA-nanodisc generated by cryoSPARC has several interesting asymmetric features (Figure 8). As mentioned previously, there are three LF$_N$ densities. A molecular dynamics flexible fitting (MDFF) of 3LF$_N$-PA pore docked three LF$_N$, in pink, magenta, and purple, in between subunit interfaces of PA, as was seen previously in the prepore crystal structure of 4LF$_N$-8PA and confirmed by 15 Å cryo-EM structures using the complete LF-PA prepore structure [16,20]. Previous work has shown the N-terminal tail of LF$_N$ feeds into the pore lumen and interacts with the Phe clamp. A cross-section of the model, shown in Figure 9, reveals the narrowing of the pore lumen is consistent with the positioning of the Phe clamp region in the MDFF model. Curiously, this pH 7.5 low-resolution triply-bound LF$_N$-PA pore structure shows an open pore region, in contrast to the closed densities observed for the previous lower-resolution, seven-fold symmetrized structures [22].

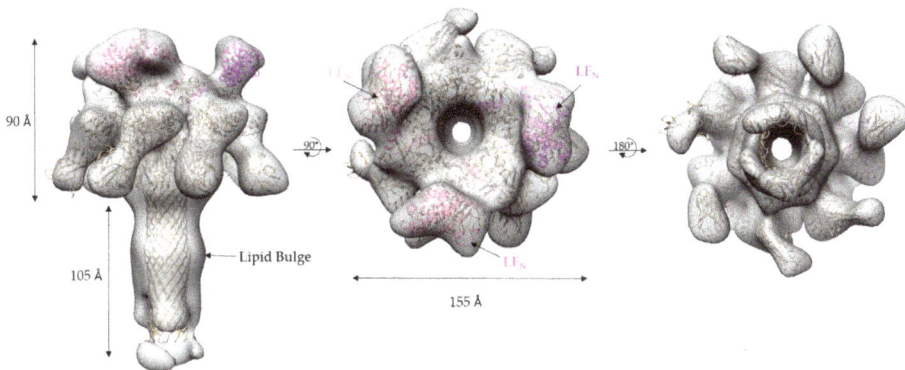

Figure 8. 3LF$_N$-PA cryo-EM density map (grey) with the ribbon structure MDFF-fitted 3LF$_N$ (pink, magenta, and purple)-PA pore (gold): (**left**) side view; (**middle**) top view; and (**right**) bottom view.

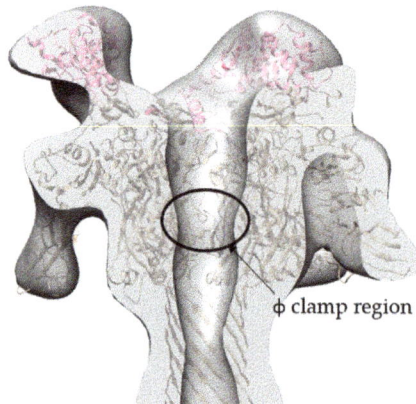

Figure 9. Cross-section of the side view cryo-EM density map (**grey**) with ribbon structure MDFF model of LF_N (**pink and magenta**) and PA (**gold**) reveals that the narrowing of the pore lumen in the density map is consistent with location of Phe clamp region.

A comparison of the MDFF atomic structure filtered to 17 Å with the 17 Å cryo-EM-derived $3LF_N$-PA pore structure showed surface details that were visually indistinguishable (Figure 10). For example, the top view of the cryo-EM $3LF_N$-PA structure showed LF_N has a distinctive bean shape (Figure 10A). A top view of the space filled PDB structure of LF_N bound to the prepore structure also had this same characteristic shape [16]. A small protrusion from the PA pore cap where LF_N is absent was also present in both models. Unlike the MDFF structure, the domain 4 regions of the cryo-EM derived structure are not equal in density, suggesting that these regions are dynamic structures as was previously observed by Jiang et al. [17]. It is also important to note that not all surface regions in the cryo-EM reconstruction are filled by MDFF analysis. For example, the β-barrel bulge that is due to lipid binding is not revealed in the fit structure since such a bulge in the highly-stable β-barrel is energetically restrictive.

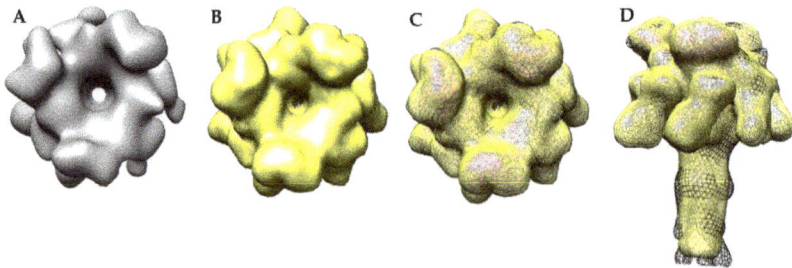

Figure 10. Comparison of the cryo-EM and MDFF models show similar topology of the LF_N bean shape and PA cap protrusion: (**A**) top view of the 17 Å cryo-EM map; (**B**) top view of the MDFF atomic resolution model filtered to 17 Å; (**C**) top view overlay of the 17 Å cryo-EM model (black mesh), the MDFF model filtered to 17 Å (yellow), and the MDFF ribbon model (PA in gold, LF_N in magenta); (**D**) side view overlay of the cryo-EM model (black mesh), the MDFF model filtered to 17 Å (yellow), and the MDFF ribbon model (PA in gold, LF_N in magenta).

3. Discussion

Atomic resolution cryo-EM is a rapidly evolving structural method that can be applied to examine the atomic consequences of LF_N interactions with the PA pore. The ability to generate soluble, lipid-stabilized LF$_N$-PA pore structures, even in this low resolution model, is the critical, important first step in demonstrating that we can obtain structural snapshots of this complex.

3.1. Sample Preparation of Highly Pure Complexes

One of the main thrusts of this work has been to demonstrate that we can routinely obtain highly-pure engagement complexes (multiply- or singly-bound LF$_N$) using an immobilization bead-based protocol and nanodisc technology without using columns to purify the final complexes [23,25] and minimizing detergent influences on structure [31,32]. Even at 17 Å resolution, the variability of the domain 4 densities for the LF$_N$-PA pore indicates this region is intrinsically flexible [17], ruling out the possibility that this flexibility is due to grid adherence constraints. Although it is possible the insertion of the tip region into an authentic lipid bilayer (e.g., a nanodisc) may result in more ordered structures, better nanodisc resolution is required to make this assessment [33–35]. Previously, protein-bilayer interactions in nanodiscs have been noted to result in extended β-barrel protein structures (approximately two residues per strand) compared with detergent-solubilized structures [35].

3.2. Initial Cryo-EM Model of 3LF$_N$-PA Pore

The cryo-EM density map structure was created without imposing symmetry or biasing towards an initial input model using the cryoSPARC *ab initio* reconstruction and subsequent refinement procedures. This 17 Å 3LF$_N$-PA pore model showed three distinct LF$_N$ densities. In agreement with what was observed previously, the LF$_N$ densities are positioned between two protomer interfaces of the PA pore [16,20]. The main contact points are on the crest of the pore and in the α-clamp. Only three LF$_N$ are able to bind to a heptameric pore, leaving one protomer without any direct LF$_N$ contacts.

A cross-section through the EM density map showed the location of the pore opening complete with the narrowing of the pore lumen. An MDFF fit starting from the atomic resolution pore structure with LF$_N$ bound positions this narrowing region with the Phe clamp loop region and preserves the opening at the Phe clamp annulus. While the number of particles and subsequent resolution of this current cryo-EM density map do not allow us to definitively define structural details of the pore lumen, it would be of interest to determine if the pore remains in a more open configuration at pH 7.5 when one or three LF monomers are bound. This further highlights the need to obtain high-resolution structures of the PA pore with one or more LF bound to determine if the Phe clamp region remains more open under these conditions. As mentioned previously, the presence of interfering electrostatic interactions appears to lead to a more open pore structure. Notably, this open pore diameter has been suggested by Das and Krantz to be necessary in order to accommodate α-helical regions during translocation at pH 5.0. These atomic resolution structures will be key to determining if varying ratios of LF bound (i.e., one vs. three) induces significant structural asymmetry (variable positioning of the Phe clamp) or concerted symmetry (all open) on the PA pore structure.

It is not uncommon to observe both small- and large-scale symmetry breakage of ordered oligomers induced by protein-protein interactions. For example, structures of protein substrate and nucleotide interactions with GroEL, a tetradecameric ring chaperonin protein, show very discernable asymmetric adjustments due to protein substrate interactions [36,37], as well as ATP binding and hydrolysis [38]. A more dramatic demonstration for ligand-induced distortion of symmetry is observed for the ATP bound vs. ADP bound ATPase unfolding machinery of the valosin-containing protein-like ATPase (VAT) recently resolved by cryo-EM [39]. In this instance, the hexameric structure was dramatically distorted in the presence of ADP and appeared to coincide with its ATP/ADP conformational switching mechanism to provide a conformational platform that unfolds proteins prior to degradation.

It would be of great interest to compare singly bound and multiply bound LF$_N$-PA pore structures in different pH conditions in order to discern any distinct structural differences that may result from being in various pH environments. Observing these different states of the engagement complex (pH 5.0 vs. pH 7.5, 1 LF$_N$ vs. 3 LF$_N$) would be useful in determining the position of the Phe clamp loop region and potentially defining unstructured regions of the LF$_N$ that may become structured upon binding to the pore prior to translocation at pH 5.0. There are existing crosslinking studies by the Collier group indicating this interaction is present at pH 5.0 [13]. Thus, there is precedence for this interaction and those cryo-EM structure collection experiments at pH 5.0 are currently underway. In all cases, given the intrinsic stability of the extended β-barrel at pH 5.0 and pH 7.0, it is highly unlikely that the β-barrel region will be structurally altered when LF$_N$ binds to the PA pore cap region. Rather, the more flexible parts of the PA pore (i.e., the cap region, Phe clamp region, etc.) will be highly susceptible to LF$_N$-induced conformational changes. How LF structurally impacts translocation and pore formation may be manifested through long range allosteric affects.

4. Conclusions

Understanding both PA pore formation and LF translocation through the PA pore is crucial to mitigating, and perhaps preventing, anthrax disease. To better understand the interactions between LF$_N$ and the PA pore, the structure of LF$_N$-bound PA pore was examined using cryo-EM. The 17 Å structure of PA pore with 3 LF$_N$ bound was the result of pore immobilization, nanodisc solubilization, *ab initio* modeling, and refinement. In this pH 7.5 structure, the contributions from the three unstructured LF$_N$ lysine-rich tail regions do not occlude the Phe clamp opening, indicating these flexible tails remain unstructured and unresolved. The next structures to examine are the LF$_N$-PA pore complexes at pH 5.0 to determine if the unstructured LF N-terminal tails interact with the Phe clamp.

5. Materials and Methods

5.1. Protein Expression and Purification

Recombinant wild-type (WT) PA was expressed in the periplasm of *Escherichia coli* BL21 (DE3) and purified by anion exchange chromatography [40] after activation of PA with trypsin [41]. QuikChange site-directed mutagenesis (Stratagene) was used to introduce mutations into the plasmid (pET SUMO (Invitrogen)) encoding a truncated recombinant portion of lethal factor. LF$_N$ E126C and was expressed as His$_6$-SUMO-LF$_N$, which was later cleaved by SUMO (small ubiquitin-related modifier) protease, revealing the native LF$_N$ E126C N-terminus [41]. Membrane scaffold protein 1D1 (MSP1D1) was expressed from the pMSP1D1 plasmid (AddGene) with an N-terminal His-tag and was purified by immobilized Ni-NTA affinity chromatography as previously described [42].

5.2. Formation of LF$_N$-PA-Nanodisc Complexes

Heterogeneous LF$_N$-PA-nanodisc complexes were formed and purified as previously described [22,25]. In brief, E126C LF$_N$ was immobilized by coupling E126C LF$_N$ to activated thiol sepharose 4B beads (GE Healthcare Bio-Sciences, Pittsburgh, PA, USA) in Assembly Buffer (50 mM Tris, 50 mM NaCl, pH 7.5) at 4 °C for 12 h. One hundred microliters (100 μL) of 0.2 μM heptameric WT PA prepore was then added to 50 μL of LF$_N$ bead slurry. Beads were washed three times with Assembly Buffer to remove any unbound PA prepores. The immobilized LF$_N$-PA prepore complexes were then incubated in 1 M urea (Thermo Fisher Scientific, Waltham, MA, USA) at 37 °C for 5 min to transition the PA prepores to pores. After three more washes with Assembly Buffer, pre-nanodisc micelles (2.5 μM MSP1D1, 162.5 μM 1-palmitoyl-2-oleoyl-sn-glycero-3-phosphocholine (POPC) (Avanti, Alabaster, AL, USA) in 25 mM Na-cholate (Sigma-Aldrich, St. Louis, MO, USA), 50 mM Tris, and 50 mM NaCl) were added and bound to the aggregation-prone hydrophobic transmembrane β-barrel of PA. The micelles were collapsed into nanodiscs by removing Na-cholate using dialysis with Bio-Beads (BIO RAD, Hercules, CA, USA) as previously described [43]. Soluble complexes were released from the thiol sepharose beads by reducing the E126C LF$_N$-bead disulfide bond using 50 mM dithiothreitol (DTT) (Goldbio, St. Louis, MO, USA) in Assembly Buffer. To select for LF$_N$-PA-nanodisc complexes, the released complexes were then incubated with Ni-NTA resin (Qiagen, Germantown, MD, USA). The His-tag on the MSP1D1 construct bound to the resin. Complexes were eluted from the Ni-NTA using 200 mM imidazole (Sigma-Aldrich, St. Louis, MO, USA) in Assembly Buffer. Assembled complexes were initially confirmed using negative-stain TEM.

Homogeneous 1LF$_N$-PA pore complexes were produced using a modified protocol where E126C LF$_N$ and PA were incubated in solution at a ratio of 1LF$_N$:2PA prior to immobilization to reduce the number of complexes with multiple bound LF$_N$. In this particular instance, affinity purification with Ni-NTA resin was omitted to minimize sample loss and homogeneous samples were still obtained.

5.3. Cryo-EM Sample Preparation and Data Collection

Cryo-EM samples were prepared within 10–30 min of elution. Three to four microliters (3–4 μL) of purified LF$_N$-PA-nanodisc complexes were added to a glow-discharged holey carbon grid (Quantifoil R3/4 300 M Cu holey carbon) (Electron Microscopy, Sciences, Hatfield, PA, USA) and plunge frozen in liquid ethane using a Vitrobot (FEI, Hilsboro, OR, USA). Data were collected manually over the course of 10 sessions (8–10 h each) on a 2200FSC election microscope (JEOL, Peabody, MA, USA) at NCMI, Baylor College of Medicine. The microscope was equipped with an in-column energy filter (using a 20 eV slit) and operated at 200 kV acceleration voltage. Images were recorded on a Gatan 4k × 4k CCD camera using a 60,000× nominal magnification (1.81 Å/pixel) with an overall range of defocus values from one to three microns using a dose of approximately 20 e$^-$/Å^2. Approximately 650 individual micrographs were recorded. Homogeneous 1LF$_N$-PA-nanodisc complexes were imaged and screened using a Tecnai F30 G2 twin transmission electron microscope (FEI, Hillsboro, OR, USA) at 200 kV at the University of Missouri Electron Microscopy Core Facility (EMC).

5.4. Image Analysis and 3D Reconstruction

The 650 raw micrographs obtained at Baylor were evaluated using EMAN2.1 [44]. At the early evaluation stage, around 250 of these micrographs were rejected due to either gross contamination or charging artifacts visible in the Fourier transforms. A total of 30,696 particles were manually boxed out using the e2boxer.py routine of EMAN2.1 with a box size of 224 × 224 pixels. The data evaluated with EMAN2.1 and RELION, showed a heterogeneous population of single, double, and triple LF$_N$-bound PA. Due to this heterogeneity, it was difficult to use earlier versions of RELION with this smaller dataset to produce a model without imposing C7 symmetry. The approximately 30,000 particles were reevaluated using cryoSPARC (version 0.5). First, 2D class averaging was performed (Figure 5A). Bad classes were visually identified and discarded (e.g., unrecognizable densities, smaller than predicted density envelopes, etc.). Using the remaining 18,806 good particles, an *ab initio* reconstruction using the cryoSPARC SGD was carried out to computationally purify the dataset into subsets containing one, two, or three bound LF$_N$. This computation was performed with the following settings: four groups, a group similarity factor of 0.2, and 10-fold the default iterations.

The SGD algorithm allows for *ab initio* structure determination that is insensitive to initial model inputs. An arbitrary computer-generated random initialization model improves over many noisy model iterations. Each step is based on the gradient of the approximated objective function obtained with a random selection of a small batch of initial particles. These approximate gradients do not exactly match the "overall optimization objective" (best *ab initio* model) but through multiple rounds, the derived models gradually approach this maximum. As stated by Punjani, Brubaker, and colleagues, "the success of SGD is commonly explained by the noisy sampling approximation allowing the algorithm to widely explore the space of all 3D maps to finally arrive near the correct structure" [28,29]. In contrast to using the entire dataset for initial model reconstruction, cryoSPARC samples random subsets of the images during its rapid iteration processes.

The *ab initio* model with three clearly-resolved LF$_N$ densities possessed the largest percentage of particles (44.9%). The second most populated class (20.1%) appeared to contain one prominent LF$_N$ density with the hint of a second bound LF$_N$, but requires more particles in order to achieve definition (Figure 11B, column 1). After the *ab initio* model was generated, a homogeneous refinement with 100 additional passes using the branch-to-bound maximum likelihood optimization cryoSPARC algorithm. The final cryo-EM map resolution was estimated to be 17 Å based on Fourier Shell Correlation (FSC) with a cut off of 0.5 (Figure 5). The *ab initio* group 1 with the second highest percentage (20.1%) had one LF_N density at a lower volume threshold. However, further processing of the potential single bound LF_N revealed added density on the PA pore cap from a mixture of one and two LF_N populations (Figure 11C, column 1). More particles are needed to populate this distribution before definitive single or double LF_N-bound structures can be obtained.

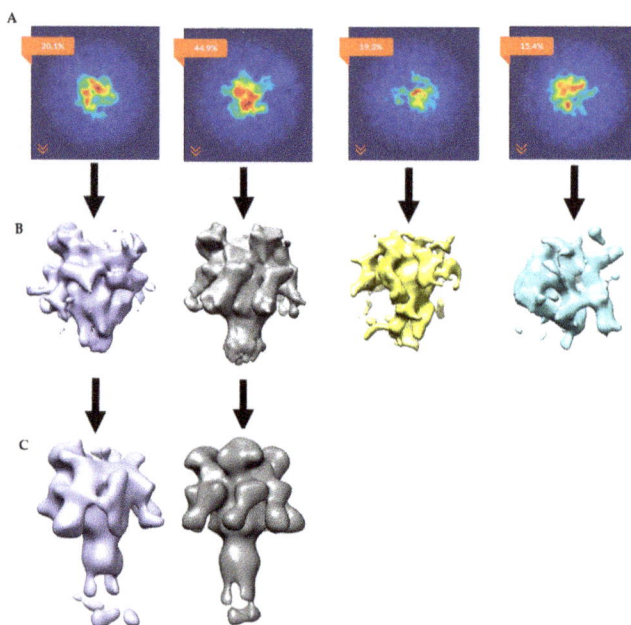

Figure 11. CryoSPARC data analyses parsed out heterogeneous LF_N-PA-nanodiscs: (**A**) Image projection of heterogeneous *ab initio* reconstruction with four groups, the largest group, with 44.9% of particles, corresponds to $3LF_N$; (**B**) *ab initio* 3D models (side views); and (**C**) homogeneous refinements of *ab initio* group 1 and group 2. Group 2 refined to 18 Å model of $3LF_N$-PA from 4732 particles. Group 1 clearly shows missing density in the cap region and will need more particles to determine if this structure contains sub-saturated populations (i.e., one or two LF_N bound) of LF_N bound to the PA pore structure or that this group will split out further to separate one vs. two LF_N-bound populations.

The cryoSPARC 3D reconstruction software tool (Structura Biotechnology, Toronto, ON, Canada) was run on a single workstation (Nova 2 Model: 2 × NVIDIA Titan Xp GPU, Intel Xeon E5-1630v4 (4-core 3.7 GHz CPU), 64 GB DDR4-2400 RAM, Intel 1.2 TB SATA solid state drive for runtime cache, and 4 × 4 TB Seagate SATA HDDs) purchased from Silicon Mechanics (Bothell, WA, USA) housed in the Fisher Laboratory. One of the main advantages of using cryoSPARC in combination with this computer system is the reduced computational time. What was once days or weeks in computational time is now only minutes or hours [29]. For example, as this paper was being written, the latest version of cryoSPARC was released (upgrade from 0.41 to 0.5). All Baylor collected data was reanalyzed with the newer version as a test for reproducibility in the span of 4 h (from reevaluating 2D classification, removing poor particles, etc.) where the final output *ab initio* models, reevaluated 2D class averages from separated populations and refined structures were reproduced using the single workstation described above. The use of SGD algorithms to generate *ab initio* models are now being beta tested or implemented in other software packages.

5.5. Molecular Dynamics Flexible Fitting of 3LF_N-PA

A molecular model was fit into the cryo-EM density map using molecular dynamics flexible fitting (MDFF) methods [45] which apply an additional potential derived from the density map to the molecules. The starting molecular model was built by rigid docking three LF_N (PDB 3KWV) onto the PA pore cap (PDB 3J9C). The cryo-EM density map and initial molecular model were spatially aligned in Sculptor [46,47]. The density map was then converted from mrc to a situs file extension for compatibility with the Visual Molecular Dynamics (VMD) software suite. The atomic model and density map files were prepared for MDFF fitting in VMD by the typical MDFF tutorial progression [47,48]. The model was minimized for 2000 steps simulated for 50 ps at 300 K in vacuum. The grid-scaling factor, which controls the relative strength of the MDFF potential was set to 0.3. Figure 10 compares the 17 Å filtered MDFF structure with the 17 Å cryo-EM derived structure to show distinct similarities in surface topologies [46,48,49].

Acknowledgments: This work was supported by NIH R01AI090085 (M.T.F.), NIH U54GM087519 (W.I.), Madison and Lila Self Graduate Fellowship (A.J.M.), NIH 5T32GM008359 (P.T.O.), KUMC Bridging Funds (C.T.). In addition, some sample preparation and data collection was performed at the National Center for Macromolecular Imaging (NCMI), supported by NIH grant NIGMS P41 GM 103832 (Wah Chiu). The authors would also like to thank John R. Collier and Brad L. Pentelute for providing the initial PA prepore and LF_N constructs as well as their helpful suggestions.

Author Contributions: M.T.F., N.A., and A.J.M. conceived and designed the experiments; A.J.M., N.A., and R.D. performed the experiments; A.J.M., N.A., S.M., Y.Q., W.I., and C.T. analyzed the data; A.J.M., P.T.O., E.P.G., T.A.W., and C.T. contributed purified proteins, reagents, materials, and/or analysis tools; A.J.M. and M.T.F. wrote the paper; A.J.M., N.A., C.T., P.T.O., W.I., E.P.G. and M.T.F. edited the paper.

Conflicts of Interest: The authors declare no conflict of interest.

References

1. Young, J.A.; Collier, R.J. Anthrax toxin: Receptor binding, internalization, pore formation, and translocation. *Annu. Rev. Biochem.* **2007**, *76*, 243–265. [CrossRef] [PubMed]
2. Lacy, D.B.; Wigelsworth, D.J.; Melnyk, R.A.; Harrison, S.C.; Collier, R.J. Structure of heptameric protective antigen bound to an anthrax toxin receptor: A role for receptor in pH-dependent pore formation. *Proc. Natl. Acad. Sci. USA* **2004**, *101*, 13147–13151. [CrossRef] [PubMed]
3. Santelli, E.; Bankston, L.A.; Leppla, S.H.; Liddington, R.C. Crystal structure of a complex between anthrax toxin and its host cell receptor. *Nature* **2004**, *430*, 905. [CrossRef] [PubMed]
4. Wimalasena, D.S.; Janowiak, B.E.; Lovell, S.; Miyagi, M.; Sun, J.; Zhou, H.; Hajduch, J.; Pooput, C.; Kirk, K.L.; Battaile, K.P. Evidence that histidine protonation of receptor-bound anthrax protective antigen is a trigger for pore formation. *Biochemistry* **2010**, *49*, 6973–6983. [CrossRef] [PubMed]
5. Mogridge, J.; Cunningham, K.; Collier, R.J. Stoichiometry of anthrax toxin complexes. *Biochemistry* **2002**, *41*, 1079–1082. [CrossRef] [PubMed]
6. Kintzer, A.F.; Thoren, K.L.; Sterling, H.J.; Dong, K.C.; Feld, G.K.; Tang, I.I.; Zhang, T.T.; Williams, E.R.; Berger, J.M.; Krantz, B.A. The protective antigen component of anthrax toxin forms functional octameric complexes. *J. Mol. Biol.* **2009**, *392*, 614–629. [CrossRef] [PubMed]
7. Krantz, B.A.; Finkelstein, A.; Collier, R.J. Protein translocation through the anthrax toxin transmembrane pore is driven by a proton gradient. *J. Mol. Biol.* **2006**, *355*, 968–979. [CrossRef] [PubMed]
8. Brown, M.J.; Thoren, K.L.; Krantz, B.A. Role of the α clamp in the protein translocation mechanism of anthrax toxin. *J. Mol. Biol.* **2015**, *427*, 3340–3349. [CrossRef] [PubMed]
9. Das, D.; Krantz, B.A. Secondary structure preferences of the anthrax toxin protective antigen translocase. *J. Mol. Biol.* **2017**, *429*, 753–762. [CrossRef] [PubMed]
10. Krantz, B.A.; Melnyk, R.A.; Zhang, S.; Juris, S.J.; Lacy, D.B.; Wu, Z.; Finkelstein, A.; Collier, R.J. A phenylalanine clamp catalyzes protein translocation through the anthrax toxin pore. *Science* **2005**, *309*, 777–781. [CrossRef] [PubMed]

11. Melnyk, R.A.; Collier, R.J. A loop network within the anthrax toxin pore positions the phenylalanine clamp in an active conformation. *Proc. Natl. Acad. Sci. USA* **2006**, *103*, 9802–9807. [CrossRef] [PubMed]

12. Janowiak, B.E.; Fischer, A.; Collier, R.J. Effects of introducing a single charged residue into the phenylalanine clamp of multimeric anthrax protective antigen. *J. Biol. Chem.* **2010**, *285*, 8130–8137. [CrossRef] [PubMed]

13. Janowiak, B.E.; Jennings-Antipov, L.D.; Collier, R.J. Cys–Cys cross-linking shows contact between the n-terminus of lethal factor and phe427 of the anthrax toxin pore. *Biochemistry* **2011**, *50*, 3512–3516. [CrossRef] [PubMed]

14. Wynia-Smith, S.L.; Brown, M.J.; Chirichella, G.; Kemalyan, G.; Krantz, B.A. Electrostatic ratchet in the protective antigen channel promotes anthrax toxin translocation. *J. Biol. Chem.* **2012**, *287*, 43753–43764. [CrossRef] [PubMed]

15. Pannifer, A.D.; Wong, T.Y.; Schwarzenbacher, R.; Renatus, M.; Petosa, C.; Bienkowska, J.; Lacy, D.B.; Collier, R.J.; Park, S.; Leppla, S.H. Crystal structure of the anthrax lethal factor. *Nature* **2001**, *414*, 229–233. [CrossRef] [PubMed]

16. Feld, G.K.; Thoren, K.L.; Kintzer, A.F.; Sterling, H.J.; Tang, I.I.; Greenberg, S.G.; Williams, E.R.; Krantz, B.A. Structural basis for the unfolding of anthrax lethal factor by protective antigen oligomers. *Nat. Struct. Mol. Biol.* **2010**, *17*, 1383–1390. [CrossRef] [PubMed]

17. Jiang, J.; Pentelute, B.L.; Collier, R.J.; Zhou, Z.H. Atomic structure of anthrax pa pore elucidates toxin translocation. *Nature* **2015**, *521*, 545. [CrossRef] [PubMed]

18. Sellman, B.R.; Nassi, S.; Collier, R.J. Point mutations in anthrax protective antigen that block translocation. *J. Biol. Chem.* **2001**, *276*, 8371–8376. [CrossRef] [PubMed]

19. Sun, J.; Lang, A.E.; Aktories, K.; Collier, R.J. Phenylalanine-427 of anthrax protective antigen functions in both pore formation and protein translocation. *Proc. Natl. Acad. Sci. USA* **2008**, *105*, 4346–4351. [CrossRef] [PubMed]

20. Fabre, L.; Santelli, E.; Mountassif, D.; Donoghue, A.; Biswas, A.; Blunck, R.; Hanein, D.; Volkmann, N.; Liddington, R.; Rouiller, I. Structure of anthrax lethal toxin prepore complex suggests a pathway for efficient cell entry. *J. Gen. Physiol.* **2016**, *148*, 313–324. [CrossRef] [PubMed]

21. Pentelute, B.L.; Sharma, O.; Collier, R.J. Chemical dissection of protein translocation through the anthrax toxin pore. *Angew. Chem. Int. Ed. Engl.* **2011**, *50*, 2294–2296. [CrossRef] [PubMed]

22. Gogol, E.; Akkaladevi, N.; Szerszen, L.; Mukherjee, S.; Chollet-Hinton, L.; Katayama, H.; Pentelute, B.; Collier, R.; Fisher, M. Three dimensional structure of the anthrax toxin translocon–lethal factor complex by cryo-electron microscopy. *Protein Sci.* **2013**, *22*, 586–594. [CrossRef] [PubMed]

23. Akkaladevi, N.; Hinton-Chollet, L.; Katayama, H.; Mitchell, J.; Szerszen, L.; Mukherjee, S.; Gogol, E.; Pentelute, B.; Collier, R.; Fisher, M. Assembly of anthrax toxin pore: Lethal-factor complexes into lipid nanodiscs. *Protein Sci.* **2013**, *22*, 492–501. [CrossRef] [PubMed]

24. Naik, S.; Brock, S.; Akkaladevi, N.; Tally, J.; Mcginn-Straub, W.; Zhang, N.; Gao, P.; Gogol, E.; Pentelute, B.; Collier, R.J. Monitoring the kinetics of the pH-driven transition of the anthrax toxin prepore to the pore by biolayer interferometry and surface plasmon resonance. *Biochemistry* **2013**, *52*, 6335–6347. [CrossRef] [PubMed]

25. Akkaladevi, N.; Mukherjee, S.; Katayama, H.; Janowiak, B.; Patel, D.; Gogol, E.P.; Pentelute, B.L.; Collier, R.J.; Fisher, M.T. Following natures lead: On the construction of membrane-inserted toxins in lipid bilayer nanodiscs. *J. Membr. Boil.* **2015**, *248*, 595–607. [CrossRef] [PubMed]

26. Katayama, H.; Wang, J.; Tama, F.; Chollet, L.; Gogol, E.; Collier, R.; Fisher, M. Three-dimensional structure of the anthrax toxin pore inserted into lipid nanodiscs and lipid vesicles. *Proc. Natl. Acad. Sci. USA* **2010**, *107*, 3453–3457. [CrossRef] [PubMed]

27. Yang, Z.; Fang, J.; Chittuluru, J.; Asturias, F.J.; Penczek, P.A. Iterative stable alignment and clustering of 2d transmission electron microscope images. *Structure* **2012**, *20*, 237–247. [CrossRef] [PubMed]

28. Brubaker, M.A.; Punjani, A.; Fleet, D.J. Building Proteins in a Day: Efficient 3D Molecular Reconstruction. In Proceedings of the IEEE Conference on Computer Vision and Pattern Recognition, Boston, MA, USA, 7–12 June 2015; pp. 3099–3108.

29. Punjani, A.; Rubinstein, J.L.; Fleet, D.J.; Brubaker, M.A. Cryosparc: Algorithms for rapid unsupervised cryo-EM structure determination. *Nat. Methods* **2017**, *14*, 290–296. [CrossRef] [PubMed]

30. Ripstein, Z.A.; Huang, R.; Augustyniak, R.; Kay, L.E.; Rubinstein, J.L. Structure of a AAA+ unfoldase in the process of unfolding substrate. *eLife* **2017**, *6*, e25754. [CrossRef] [PubMed]

31. Shen, P.S.; Yang, X.; DeCaen, P.G.; Liu, X.; Bulkley, D.; Clapham, D.E.; Cao, E. The structure of the polycystic kidney disease channel pkd2 in lipid nanodiscs. *Cell* **2016**, *167*, 763–773. [CrossRef] [PubMed]

32. Palazzo, G.; Lopez, F.; Mallardi, A. Effect of detergent concentration on the thermal stability of a membrane protein: The case study of bacterial reaction center solubilized by *N*,*N*-dimethyldodecylamine-*N*-oxide. *Biochim. Biophys. Acta* **2010**, *1804*, 137–146. [CrossRef] [PubMed]

33. Patargias, G.; Bond, P.J.; Deol, S.S.; Sansom, M.S. Molecular dynamics simulations of GlpF in a micelle vs in a bilayer: Conformational dynamics of a membrane protein as a function of environment. *J. Phys. Chem. B* **2005**, *109*, 575–582. [CrossRef] [PubMed]

34. Cox, K.; Sansom, M.S. One membrane protein, two structures and six environments: A comparative molecular dynamics simulation study of the bacterial outer membrane protein pagp. *Mol. Membr. Biol.* **2009**, *26*, 205–214. [CrossRef] [PubMed]

35. Eddy, M.T.; Su, Y.; Silvers, R.; Andreas, L.; Clark, L.; Wagner, G.; Pintacuda, G.; Emsley, L.; Griffin, R.G. Lipid bilayer-bound conformation of an integral membrane beta barrel protein by multidimensional MAS NMR. *J. Biomol. NMR* **2015**, *61*, 299–310. [CrossRef] [PubMed]

36. Elad, N.; Clare, D.K.; Saibil, H.R.; Orlova, E.V. Detection and separation of heterogeneity in molecular complexes by statistical analysis of their two-dimensional projections. *J. Struct. Biol.* **2008**, *162*, 108–120. [CrossRef] [PubMed]

37. Weaver, J.; Jiang, M.; Roth, A.; Puchalla, J.; Zhang, J.; Rye, H.S. Groel actively stimulates folding of the endogenous substrate protein pepq. *Nat. Commun.* **2017**, *8*. [CrossRef] [PubMed]

38. Saibil, H.R.; Ranson, N.A. The chaperonin folding machine. *Trends Biochem. Sci.* **2002**, *27*, 627–632. [CrossRef]

39. Huang, R.; Ripstein, Z.A.; Augustyniak, R.; Lazniewski, M.; Ginalski, K.; Kay, L.E.; Rubinstein, J.L. Unfolding the mechanism of the aaa+ unfoldase vat by a combined cryo-em, solution nmr study. *Proc. Natl. Acad. Sci. USA* **2016**, *113*, 4190–4199. [CrossRef] [PubMed]

40. Miller, C.J.; Elliott, J.L.; Collier, R.J. Anthrax protective antigen: Prepore-to-pore conversion. *Biochemistry* **1999**, *38*, 10432–10441. [CrossRef] [PubMed]

41. Wigelsworth, D.J.; Krantz, B.A.; Christensen, K.A.; Lacy, D.B.; Juris, S.J.; Collier, R.J. Binding stoichiometry and kinetics of the interaction of a human anthrax toxin receptor, cmg2, with protective antigen. *J. Biol. Chem.* **2004**, *279*, 23349–23356. [CrossRef] [PubMed]

42. Ritchie, T.; Grinkova, Y.; Bayburt, T.; Denisov, I.; Zolnerciks, J.; Atkins, W.; Sligar, S. Chapter eleven-reconstitution of membrane proteins in phospholipid bilayer nanodiscs. *Methods Enzymol.* **2009**, *464*, 211–231. [PubMed]

43. Denisov, I.; Grinkova, Y.; Lazarides, A.; Sligar, S. Directed self-assembly of monodisperse phospholipid bilayer nanodiscs with controlled size. *J. Am. Chem. Soc.* **2004**, *126*, 3477–3487. [CrossRef] [PubMed]

44. Tang, G.; Peng, L.; Baldwin, P.R.; Mann, D.S.; Jiang, W.; Rees, I.; Ludtke, S.J. Eman2: An extensible image processing suite for electron microscopy. *J. Struct. Biol.* **2007**, *157*, 38–46. [CrossRef] [PubMed]

45. Qi, Y.; Lee, J.; Singharoy, A.; McGreevy, R.; Schulten, K.; Im, W. Charmm-gui mdff/xmdff utilizer for molecular dynamics flexible fitting simulations in various environments. *J. Phys. Chem. B* **2016**, *121*, 3718–3723. [CrossRef] [PubMed]

46. Birmanns, S.; Rusu, M.; Wriggers, W. Using sculptor and situs for simultaneous assembly of atomic components into low-resolution shapes. *J. Struct. Biol.* **2011**, *173*, 428–435. [CrossRef] [PubMed]

47. Wahle, M.; Wriggers, W. Multi-scale visualization of molecular architecture using real-time ambient occlusion in sculptor. *PLoS Comput. Biol.* **2015**, *11*, e1004516. [CrossRef] [PubMed]

48. Humphrey, W.; Dalke, A.; Schulten, K. Vmd: Visual molecular dynamics. *J. Mol. Graph.* **1996**, *14*, 33–38. [CrossRef]

49. Trabuco, L.G.; Villa, E.; Mitra, K.; Frank, J.; Schulten, K. Flexible fitting of atomic structures into electron microscopy maps using molecular dynamics. *Structure* **2008**, *16*, 673–683. [CrossRef] [PubMed]

toxins

MDPI

Article

Cellular Entry of the Diphtheria Toxin Does Not Require the Formation of the Open-Channel State by Its Translocation Domain

Alexey S. Ladokhin [1,]*, **Mauricio Vargas-Uribe** [1], **Mykola V. Rodnin** [1], **Chiranjib Ghatak** [1] and **Onkar Sharma** [2]

[1] Department of Biochemistry and Molecular Biology, University of Kansas Medical Center, Kansas City, KS 66160, USA; mauricio.vargasu@gmail.com (M.V.-U.); mrodnin@kumc.edu (M.V.R.); c.ghatak79@gmail.com (C.G.)

[2] Department of Microbiology and Immunobiology, Harvard Medical School, Boston, MA 02115, USA; Onkar_Sharma@hms.harvard.edu

*   Correspondence: aladokhin@kumc.edu; Tel.: +1-913-588-0489

Academic Editors: Emmanuel Lemichez and Michel R. Popoff

Received: 31 August 2017; Accepted: 20 September 2017; Published: 22 September 2017

Abstract: Cellular entry of diphtheria toxin is a multistage process involving receptor targeting, endocytosis, and translocation of the catalytic domain across the endosomal membrane into the cytosol. The latter is ensured by the translocation (T) domain of the toxin, capable of undergoing conformational refolding and membrane insertion in response to the acidification of the endosomal environment. While numerous now classical studies have demonstrated the formation of an ion-conducting conformation—the Open-Channel State (OCS)—as the final step of the refolding pathway, it remains unclear whether this channel constitutes an in vivo translocation pathway or is a byproduct of the translocation. To address this question, we measure functional activity of known OCS-blocking mutants with H-to-Q replacements of *C*-terminal histidines of the *T*-domain. We also test the ability of these mutants to translocate their own *N*-terminus across lipid bilayers of model vesicles. The results of both experiments indicate that translocation activity does not correlate with previously published OCS activity. Finally, we determined the topology of TH5 helix in membrane-inserted *T*-domain using W281 fluorescence and its depth-dependent quenching by brominated lipids. Our results indicate that while TH5 becomes a transbilayer helix in a wild-type protein, it fails to insert in the case of the OCS-blocking mutant H322Q. We conclude that the formation of the OCS is not necessary for the functional translocation by the *T*-domain, at least in the histidine-replacement mutants, suggesting that the OCS is unlikely to constitute a translocation pathway for the cellular entry of diphtheria toxin in vivo.

Keywords: membrane translocation; pH-dependent refolding; bilayer insertion; tryptophan fluorescence; depth-dependent fluorescence quenching

1. Introduction

The diphtheria toxin enters and kills the cell through the combined action of its three domains: the catalytic domain (*C*-domain), the translocation domain (*T*-domain), and the receptor-binding domain (*R*-domain) [1]. The entry process starts with the *R*-domain recognizing the HB-EGF receptor on the host cell plasma membrane, leading to the endocytosis of the entire receptor-toxin complex. Subsequent acidification of the endosomal compartment induces the refolding of the *T*-domain and its insertion into the lipid bilayer. The latter causes the translocation of the *C*-domain across the endosomal membrane, which finally inhibits protein synthesis in the cytosol. Deciphering the missing molecular details of diphtheria toxin's cellular entry is relevant for understanding the entry of other toxins [2] and for the development of biomedical applications of targeted drug delivery. Indeed, diphtheria toxin has already been utilized as a prospective anti-cancer agent for the targeted delivery of cytotoxic therapy to cancer cells [3–16]. Normally, the targeted delivery is achieved by deleting the cell receptor-binding *R*-domain and combining the remaining portion (containing *T*- and *C*-domains) with proteins that selectively bind to the surface of cancer cells. Even a "receptorless" toxin (i.e., without the *R*-domain) is cytotoxic to a variety of cancer cell lines [3]. Because cancerous cells are known to produce a slightly acidic extracellular environment, we pursue the idea that targeting of "receptorless" toxin is assured by pH-triggered membrane insertion of the *T*-domain similar to that of the *pH-Low Insertion Peptide* (pHLIP) [17–26]. The previously demonstrated ability of the isolated *T*-domain to translocate relatively large macromolecules [27] in a pH-dependent manner makes it a potential candidate for targeting tumors.

In recent years, substantial progress has been made in characterizing structural, thermodynamic, and kinetic aspects of diphtheria toxin *T*-domain's interactions with lipid bilayers. As a result of these studies (Reviewed in [28]), we now understand that the insertion/refolding process occurs along a complex pathway, which includes multiple intermediates with staggered pH-dependent transitions [29–33]. A number of critical titratable residues involved in acid-induced conformational switching have been identified in a series of spectroscopic studies in vitro (e.g., H223 and H257 [32–35], E349 and D352 [36], and the C-terminal histidines H322, H323, and H372 [35,37,38]). Two recent studies of *T*-domain conductivity in planar lipid bilayers provided better understanding of the topological arrangements of the *T*-domain at the late stages of the pathway, at the so-called Open-Channel State (OCS) [39,40]. However, the central puzzle of the *T*-domain action, namely the exact mechanism of translocation, remains unresolved. One of the key aspects of the puzzle was formulated over 30 years ago by Donovan et al. [41]: is the OCS of the *T*-domain a translocation passageway (illustrated in Figure 1 by Pathway 2) or a byproduct of translocation (Pathway 1)? In this study, we present functional and spectroscopic evidence that suggest the latter option as the preferred one.

2. Results and Discussion

2.1. Comparing the Two Translocation Pathways

We illustrate two alternative pathways of translocation of the catalytic domain of diphtheria by the translocation domain in the scheme in Figure 1. The starting structure at the top is the crystallographic structure of the toxin at neutral pH [42]. The two cartoons at the bottom represent the arrangement of the toxin upon the insertion of the *T*-domain into the lipid bilayer at acidic pH, with the *R*-domain remaining on one side of the membrane (green shape) and catalytic domain translocated across the bilayer along with the *N*-terminal helices TH1-4 of the *T*-domain (red circle, residues 1–273). On the right, the hydrophobic core of the *T*-domain is drawn in the OCS conformation with the three transmembrane helices, TH5, TH8, and TH9, and a dipped hairpin formed by shorter helices TH6 and TH7. This topology is based on a combination of measurements of channel activity in planar bilayers [39,40,43] and is supported by additional spectroscopic evidence in model lipid vesicles ([30,44] and Kyrychenko et al., Journal of Membrane Biology, in press). The translocation Pathway 2 suggests that OCS is a functionally relevant state, which provides a passageway for translocation of the presumably unfolded catalytic domain to the other side of the bilayer (red circle—before translocation; pink circle—after translocation). In contrast, Pathway 1 indicates that the OCS is formed after the translocation, and passageway through the bilayer is formed via an unknown, possibly transient conformation. In order to distinguish between the two pathways, we will use new and published data to compare how various mutations affect the following measures of *T*-domain activity: (a) conductance in planar bilayers (i.e., formation of the OCS) [37,38,45–48], (b) *N*-terminus translocation in vesicles [37,49], and ultimately (c) cell death assay based on inhibition of protein synthesis in vivo [37,46,48,50].

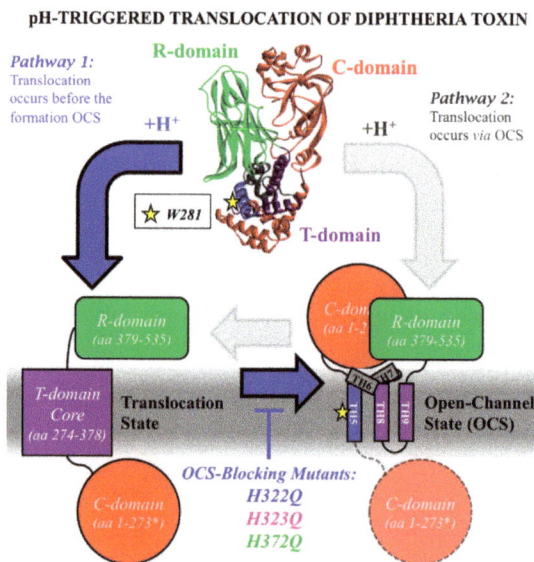

Figure 1. Schematic representation of two possible pathways for the translocation of the catalytic (C) domain across the lipid bilayer and formation of the Open-Channel State (OCS) by the translocation (T) domain of the diphtheria toxin. The starting structure on top corresponds to the crystal structure of the toxin at neutral pH [42], and consists of the C-domain (red), T-domain (helices color-coded according to OCS topology), and R-domain (green). The yellow star represents the position of W281 within the lipid bilayer in the OCS, which was used in this study to monitor the formation of the OCS (Figures 2 and 3). In pathway 1 (blue arrows), diphtheria toxin first translocates the C-domain and the T-domain's N-terminus across the lipid bilayer by an unknown mechanism (bottom left cartoon). The formation of the Open-Channel State (bottom right cartoon) is a consequence of the translocation step. In pathway 2 (gray arrows), the T-domain first adopts the OCS conformation, which then serves as passageway for the translocation of the C-domain and the T-domain's N-terminal helices. The topology of the T-domain in the OCS, i.e., transmembrane helices TH5 (blue helix), and TH8-TH9 (purple helical hairpin), and interfacial helices TH6 and TH7 (gray helices), is based on conductance measurements on planar bilayers [39,40,43] and supported by spectroscopic data in lipid vesicles [30,44]. Replacement of the C-terminal histidines of the T-domain (H322Q, H323Q, H372Q) are known to strongly reduce formation of channels in planar bilayers by blocking the formation of the OCS conformation [37,38]. Because these replacements do not affect the translocation of the N-terminus of the isolated T-domain in vitro (Figure 4a), nor delivery of the catalytic domain in vivo (Figure 4b), we conclude that the formation of the OCS is not necessary for cellular entry of the diphtheria toxin, which is occurring via Pathway 1.

2.2. Spectroscopic Evidence for the Difference in TH5 Topology in WT and in OCS-Blocking Mutant H322Q

We used tryptophan fluorescence to explore the effect of the C-terminal histidine replacement in H322Q mutant in the pH-triggered conversion of the T-domain into an OCS-like conformation in model lipid vesicles. The transmembrane positioning of TH5 helix is a hallmark of the OCS structure [40,43], making W281 located in its middle a convenient spectroscopic probe for membrane insertion. Previously, we have (a) reported that replacement of C-terminal histidines (especially H322) has a strong inhibitory effect on OCS activity and (b) shown a correlation between the tryptophan emission signal in vesicles and formation of the OCS in planar bilayers [38]. The reported spectral red-shift observed with OCS-blocking mutants is consistent with lack of insertion of TH5. However, the presence of two tryptophans complicates the interpretation. Now, we reexamine the spectral properties of the *T*-domain in the context of the single W281 by replacing the other tryptophan residue, W206, with a tyrosine in both the WT *T*-domain (from now on referred to as WT-like protein) and the H322Q OCS-blocking mutant. Our data in Figure 2 indicate that in the folded soluble state, at neutral pH, W281 is in an identical environment in both proteins. Lowering the pH below 6.5 in the presence of lipid vesicles results in a very different behavior. While in the WT-like protein, the emission maximum shifts toward shorter wavelengths, consistent with membrane insertion; in H322Q, the shift is toward longer wavelengths. The resulting 5 nm difference in position of fluorescence maximum indicates a much more hydrophobic environment of W281 in the native protein compared to that in H322Q, supporting a difference in membrane topology of TH5 helix in both proteins.

We emphasize that because of the very nature of fluorescence methodology, there can be no one-to-one link between spectral position and structure. Our data prove that known OCS-blocking mutation also results in alteration of bilayer insertion of TH5, reported by environment-sensitive shift of W281 fluorescence. Previously, we reported similar shifts for the WT protein [38], but in that case, it was impossible to specifically assign those to TH5 because of the presence of second fluorophore, W206. The results obtained with W206Y mutant (Figure 2) confirm our previous conclusion that OCS-blocking mutation H322Q alters bilayer insertion of TH5.

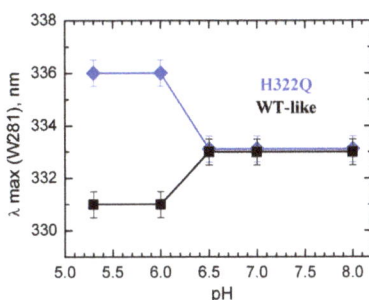

Figure 2. pH-triggered formation of the OCS conformation studied by tryptophan fluorescence. Transmembrane insertion of TH5 is monitored by measuring the position of maximum of emission (λ_{max}) of W281 in mutants of the *T*-domain carrying the W206Y replacement. The WT-like protein (black) and the mutant H322Q (blue) show different spectral behavior upon acidification of the medium in the presence of lipid vesicles. Blue-shift in the case of the WT-like protein suggests transbilayer insertion of TH5, while red-shift in the case of the mutant H322Q indicates an alternative topology of this helix.

To further investigate membrane penetration of W281, we used depth-dependent fluorescence quenching experiments [51–53]. Specifically, we measured fluorescence intensity of the membrane-inserted *T*-domain in the absence and presence of bromine atoms attached to specific sites of the lipid acyl chains. In this series, we used all three commercially available bromolipids, each with two Br atoms attached either at positions 6 and 7, 9 and 10, or 11 and 12, and plotted the quenching efficiency versus the independently determined average depth of the atoms in the bilayer [54] (Figure 3). The higher the quenching efficiency with a particular bromolipid, the closer is the transverse position of the fluorophore (W281) to that of the quencher (Br). Our data clearly indicate that the rank order of quenching efficiency of the W281 is different for the two proteins: the deeper the quencher the stronger the quenching for the WT-like protein (black squares) and the shallower the quencher the higher the quenching efficiency for H322Q mutant (blue squares).

For quantitative analysis of the data, we applied a Distribution Analysis (DA) method, which describes the quenching profile with a sum of two symmetrical Gaussian functions, representing cis-leaflet and trans-leaflet quenching [52,55]. Generally, DA uses three independent fitting parameters, one for the average depth of the fluorophore (h_m), one for the width of the fluorophore's transverse distribution (δ), and one for quenching efficiency (S), related to fluorophore's exposure to the lipid phase [56]. In this case, because the profiles are poorly resolved (i.e., maxima lie outside of the range of depths probed by quenchers), we had to use only two and fixed the δ to the value of 5.0 Å, roughly corresponding to the average value for such experiments. This is a commonly used simplifying approach [52,57], and the small variation of the value of δ does not change the overall result but affects only the quality of the fit (not shown). We estimated average depths of 4.5 ± 1.5 Å and 12.1 ± 0.1 Å from the center of the bilayer for the WT-like protein and H322Q, respectively. These values indicate a deep membrane penetration of W281 in the case of the WT-like protein and a shallow location of the same residue for the H322Q mutant. The suggested topology of TH5 in both cases, also supported by position of maximum emission (Figure 2), is drawn in Figure 3b, with TH5 as a transbilayer segment for the WT-like toxin and an interfacial helix for the mutant. We suggest that the mutation H322Q impairs the proper insertion of TH5, which is needed for the formation of the OCS.

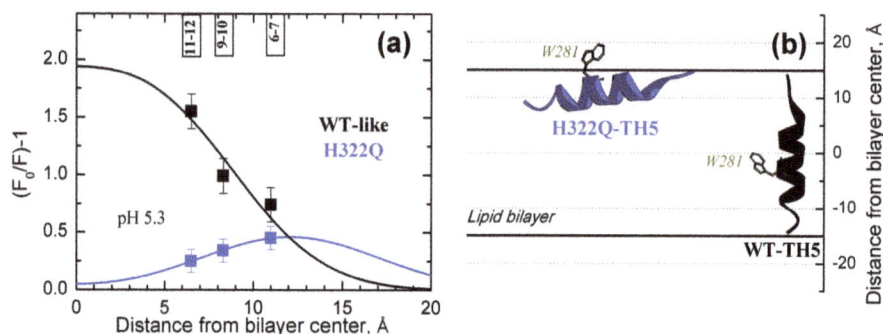

Figure 3. Depth-dependent fluorescence quenching measurements of W281 in WT-like (black) and H322Q *T*-domain (blue) using brominated lipids (both *T*-domain constructs carry W206Y mutation). (a) Quantitative analysis of penetration of W281 into the lipid vesicles containing brominated lipids. Relative quenching efficiency, F_0/F-1 (where F-fluorescence measured in the presence of bromolipids and F_0-fluorescence without bromolipids) is plotted against the average distances of bromine atoms from the center of the bilayer (11.0 Å, 8.3 Å, and 6.5 Å for 6-7-dibromo-PC, 9-10-dibromo-PC, and 11-12-dibromo-PC, respectively [54]). Solid lines represent results of fitting with a simplified version of Distribution Analysis (Equation (1)) with the following parameters: h_m = 4.5 ± 1.5 Å, S = 1.5 ± 0.3, and δ = 5.0 Å(fixed) for WT-like, and h_m = 12.1 ± 0.1 Å, S = 0.46 ± 0.1, and δ = 5.0 Å(fixed) for H322Q. (b) Schematic illustration of the proposed topology of TH5 helix in the lipid bilayer for *T*-domain WT and H322Q, consistent with positioning W281 in accordance with the data in panel (a). The TH5 adopts transbilayer topology in the OCS conformation for the WT *T*-domain, and interfacial topology for the OCS-blocking H332Q mutant.

2.3. Translocation Activity of OCS-Blocking Mutants of the T-Domain

We examined the ability of the various mutants of the *T*-domain to bridge the lipid bilayer using a cleavage-based assay developed previously [49]. Briefly, lipid vesicles are preloaded with thrombin, and then a *T*-domain containing a 17-residue thrombin-cleavable tag at the *N*-terminus is added to the external compartment. If *N*-terminus's translocation occurs upon acidification, the thrombin-cleavable tag enters the vesicle and is cleaved by thrombin. Because this changes the electrophoretic mobility of the tagged *T*-domain, *N*-terminus translocation can be detected and quantified through SDS-PAGE. In Figure 4a we show the relative translocation of the *T*-domain WT and various *C*-terminal single mutants at pH 5.8 plotted against previously determined OCS activity of the same mutants [38] (in order to allow a direct comparison with already published results, all proteins used in Figure 4 contain native W206). The measurements show that the three mutants maintain over 90% of the translocation activity despite the loss of channel activity, suggesting that replacing the OCS-blocking mutations does not alter *N*-terminus's translocation.

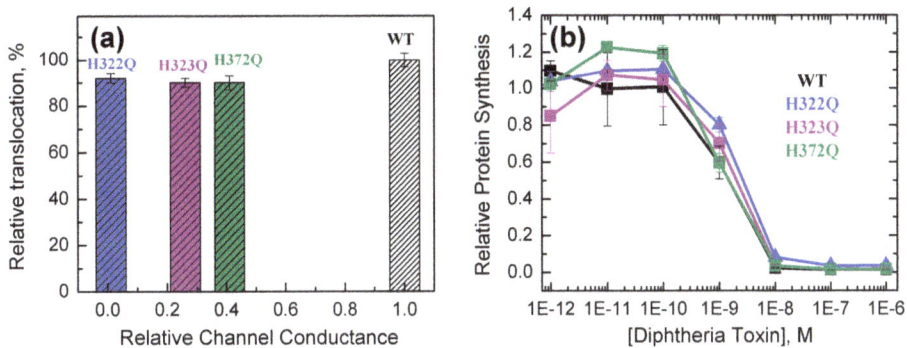

Figure 4. Activity of *T*-domain WT and mutants with single replacements in the *C*-terminal histidines. (**a**) *N*-terminus translocation in mutants of the *T*-domain relative to *N*-terminus translocation of the WT protein at pH 5.8. The assay is based on proteolytic cleavage of *N*-terminal segment preloaded into lipid vesicles followed by SDS-PAGE [49]. The data are normalized to the WT and are plotted against previously published OCS activity [38]. (**b**) Inhibition of protein synthesis by the full-length diphtheria toxin WT and the indicated mutants in CHO-K1 cells after 24 h of intoxication. Protein synthesis measurements were performed as previously detailed [37,58].

Finally, we have examined the ability of the *T*-domain mutants with His replacements to translocate the catalytic domain across the membrane using a protein synthesis inhibition assay [37,58]. Briefly, we intoxicated plated CHO-K1 cells with serial dilutions of full-length diphtheria toxin, WT or mutant and then monitored protein synthesis by measuring incorporation of L-[4,5-^{3}H]-leucine into cells after 24 h of intoxication. In Figure 4b, we show a representative result of a dose-dependent inhibition of protein synthesis by diphtheria toxin WT and the *T*-domain's mutants with *C*-terminal histidine replacements. The data indicate that both the WT and the mutants completely inhibit incorporation of radiolabeled leucine into cells and do so with similar potency. Because translocation of the *C*-domain is required for inhibition of protein synthesis, this result indirectly shows that replacing these histidines with glutamines does not affect the ability of the *T*-domain to translocate the *C*-domain. Together with the translocation assay (Figure 4a), our data demonstrate that the *T*-domain is capable of translocating its own *N*-terminus and the attached catalytic domain regardless of the OCS formation.

2.4. OCS: Critical Intermediate or Byproduct of Translocation

We answer the question in the title of this section by comparing the in vitro OCS activity of the *T*-domain in planar bilayers and cytotoxic activity in vivo for a series of mutants of diphtheria toxin with substitutions in the *T*-domain (Figure 5). The data are taken from the literature [37,38,46,48] and from Figure 4b for H322Q, H323Q, and H372Q mutants. The relative cytotoxic activity is defined as a ratio of the concentration of mutant toxin causing half-inhibition of protein synthesis (e.g., data in Figure 4b) to that for the WT toxin. Obviously, if certain mutations have little effect on either in vivo or in vitro activity, the data will cluster close to the point with coordinates 1; 1, corresponding to the WT. In contrast, if both activities are strongly disrupted, the data should cluster around the 0; 0 point. Indeed, many mutants fall into these two categories. The most interesting mutants are those that exhibit intermediate behavior, as they allow one to differentiate the limiting aspect of translocation and distinguish the two pathways outlined in Figure 1. If the formation of OCS constituted the critical intermediate step of the pathway (i.e., pathway 2), one would expect the data to follow a somewhat concave correlation pattern, illustrated by the grey arrow in Figure 5. In reality, the data clearly follow a strongly convex pattern consistent with pathway 1. In the latter pathway, the OCS is a byproduct of translocation that is or is not formed after the translocation occurs. This view is strongly supported by the high WT-like activity of OCS-blocking mutants H322Q, H323Q, and H372Q observed in both in vitro *N*-terminus translocation assay (Figure 4a) and in vivo *C*-domain translocation assay, required for cytotoxicity (Figure 4b).

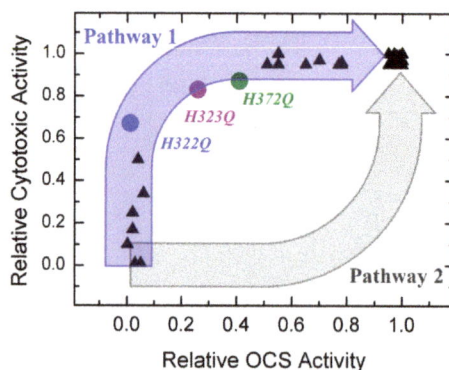

Figure 5. Comparison of the relative cellular toxicity in various mutants of diphtheria toxin and relative OCS activity of the *T*-domain with the same mutations in planar bilayers. Black triangle represents a single set of activity measurements (protein synthesis inhibition in living cells and conductance in planar bilayers) for a mutant reported in references [37,46,48]. The data for H322Q, H323Q, and H372Q mutants (color-coded circles) are from Figure 4b and reference [38]. Blue and gray arrows indicate the expected correlation for the translocation occurring via Pathway 1 or 2, respectively (see Figure 1 and text for details).

3. Conclusions and Perspectives

The data presented here clearly demonstrate the ability of diphtheria toxin with OCS-blocking mutations (e.g., H322Q, H323Q, and H372Q [38]) to ensure efficient translocation of its catalytic moiety into the cell (Figure 4b). These mutants are critical in establishing the correlation pattern between in vitro and in vivo activities of toxin mutants (Figure 5), which strongly favors Pathway 1 of cellular entry (Figure 1), in which the OCS is a byproduct of translocation rather than translocation intermediate. These results raise several questions:

1. Is it possible that these mutants take advantage, for some reason, of an entry pathway alternative to that of the WT toxin? While such an option is possible, it seems rather unlikely, because these mutants also appear active in a simplified in vitro translocation assay performed in a reductionist system of artificial lipid vesicles without a transbilayer electrical potential (Figure 4a);
2. Is it possible that the number of molecules in the OCS conformation is only a minor fraction of the entire population in model experiments? Our spectroscopic data indicate otherwise, suggesting a clear correlation between the ability of helix TH5 to insert in the OCS conformation or its precursor even in the absence of transbilayer potential (Figures 2 and 3 and [38]);
3. What is the mechanism of the translocation? Clearly more model and cellular studies will be necessary to fully answer this question. One possibility may involve the formation of a transient passageway due to the perturbation caused by the *T*-domain refolding on bilayer interface. Deciphering the molecular mechanism of this enigmatic system is especially important in light of the potential use for diphtheria toxin *T*-domain as a molecular vehicle for targeted drug delivery.

4. Materials and Methods

Materials: Palmitoyl-oleoyl-phosphatidylcholine (POPC) and Palmitoyl-oleoyl-phosphatidyl-glycerol (POPG), 6-7-dibromo-PC, 9-10-dibromo-PC, 11-12-dibromo-PC were obtained from Avanti Polar Lipids (Alabaster, AL, USA).

T-domain expression and purification: The diphtheria toxin *T*-domain (amino acids 202–378) was cloned into the NdeI- and EcoRI-treated pET15b vector. The mutations were introduced using Site-Directed Mutagenesis Kit (Stratagene, Santa Clara, CA, USA) according to manufacturer recommendations. WT *T*-domain and the mutants were expressed and purified as described previously [37,38]. Briefly, *E. coli* BL23DELysS cells previously transformed with plasmid carrying T domain WT or mutant were grown in LB medium to OD_{600} = 0.6. Protein expression was induced by addition of 0.8 mM IPTG and grown at 24 °C for 16 h. Cells were cleared by centrifugation, and pellets were resuspended in lysis buffer (25 mM Tris-HCl, 300 mM NaCl, 5 mM imidazole, lysozyme 0.1 mg/mL (Fisher Scientific, Pittsburgh, PA, USA), protease inhibitors cocktail 1× (Hoffmann-La Roche, Basel, Switzerland), pH 8) and subjected 5 times to 30 s of sonication on ice. Lysates were centrifuged at 5000 *g*, cell debris discarded, and soluble T domain was bound to Ni-NTA (Qiagen, Boston, MA, USA) at 4 °C for 16 h. Bound protein was washed several times with binding buffer (25 mM Tris-HCl, 300 mM NaCl, 5 mM imidazole) and eluted with 0.5 M imidazole in the binding buffer. Eluted protein was passed through a Superose 12 column 1 × 30 cm, flow rate 0.4 mL/min in 50 mM sodium-phosphate buffer, pH 8. Purified protein was quantified by absorbance at 280 nm (ε_{280} = 17.000 $M^{-1}cm^{-1}$) and purity was confirmed by SDS-PAGE.

On the request from editors to justify our mutagenesis strategy, we state that we do not believe phenylalanine to be a proper replacement for histidine, especially in case of membrane-interacting proteins, as it introduces additional non-natural mode of protein interaction with the lipid bilayer. For more discussion see [32].

Vesicle Preparation: Large unilamellar vesicles of diameter 0.1 μm were prepared by extrusion method [59,60] using a 1:3 molar mixture of POPC and POPG. Vesicles for quenching experiments contained 1:1 molar mixture of POPG and either 6-7-dibromo-PC, or 9-10-dibromo-PC, or 11-12-dibromo-PC, or POPC (control measurements with no quenching).

T-domain *N*-Terminus Translocation Assay: The translocation was performed using a proteolytic assay described previously [49]. Briefly, lipid vesicles composed of a molar ratio of POPG:POPC 3:1 were preloaded with ~0.02 units of bovine thrombin (Fisher Bioreagents, Fair Lawn, NJ, USA), and non-encapsulated thrombin was removed by FPLC gel filtration on a Superose 12 1 × 30 cm column. Preloaded vesicles (2 mM lipid concentration) were then mixed with 0.1 μM of *T*-domain containing an *N*-terminal His-tag linked by a sequence containing thrombin cleavage site. After 1 hour of incubation at pH 5.8, the vesicles were treated with 5% of SDS and samples analyzed by SDS-PAGE to quantify cleaved and uncleaved protein. To prevent cleavage due to vesicle rupture (rather than translocation of the *N*-terminus), 0.02 units of thrombin inhibitor hirudin (Sigma, St. Louis, MO, USA) were added to the reaction mixture.

Protein Synthesis Inhibition Assay. This assay of physiological activity of diphtheria toxin utilizes a weakened strain of full-length diphtheria toxin carrying the E148S mutation, to reduce the toxic potency [61] (as per NIH guidelines). Full-length toxin was expressed and purified as previously detailed [62]. Inhibition of protein synthesis was determined according to the previously described method [58]. Briefly, CHO-K1 cells (10,000 cells/well) were intoxicated by diphtheria toxin constructs for 24 h at 37 °C, after which incorporation of L-[4,5-^{3}H] leucine was measured.

Fluorescence Measurements: Fluorescence was measured using a SPEX Fluorolog FL 3-22 steady-state fluorescence spectrometer (Jobin Yvon, Edison, NJ, USA) equipped with double-grating excitation and emission monochromators. The measurements were made at 25 °C in 2 × 10 mm cuvettes oriented perpendicular to the excitation beam. For tryptophan fluorescence measurement, excitation wavelength was 280 nm and emission spectra was recorded between 290 nm and 500 nm using excitation and emission spectral slits of 2 and 4 nm, respectively. Normally stock concentrations of vesicles and protein were mixed to achieve final concentration of 1 μM *T*-domain and 1 mM lipid in 10 mM phosphate buffer, pH 7. The insertion was achieved by rapid addition of small aliquots of 2.5 M sodium acetate-acetic buffer, to reach the desired pH. All spectra were recorded after 30 min incubation to ensure the equilibration of the sample. Correction for the background and fitting to a log-normal distribution to determine the position of spectral maximum of emission λ_{max} was performed as described previously [63]. The spectral data were averaged to whole nm value for the presentation.

Depth-Dependent Fluorescence Quenching with Brominated Lipids: Quenching study is performed by taking a series of fluorescence measurements of the *T*-domain inserted into the vesicles containing lipids with bromine atoms attached at different positions [51,52]. Here we have used 1:1 mixture of POPG and BRPCs (6-7-dibromo-PC; 9-10-dibromo-PC and 11-12-dibromo-PC; one at a time) to prepare vesicles containing 50% quenching lipid. We have determined fluorescence intensity of tryptophan $F(h)$ as a function of the known depth of the bromolipids (h). Data are usually normalized to the intensity in the absence of quenching, F_0, measured with 1:1 POPC:POPG mixture For quantitative analysis we used Distribution Analysis method, which fits the data to the following symmetrical twin Gaussian function [52]:

$$F_0/(F(h) - 1 = G(h - h_m, \sigma, S) + G(h + h_m, \sigma, S),$$

where

$$G(h - h_m, \sigma, S) = S/(\sigma\sqrt{2\pi}) \exp\{-[(h - h_m)/2\sigma]^2\} \quad (1)$$

The three parameters of this above distribution are: most probable distance of the fluorescent probe from the center of bilayer (h_m), dispersion of the transverse distribution of the probe (σ), and quenching efficiency (S). In this particular case, we fixed the value σ at 5.0 Å, which corresponds to a typical value observed in other systems, and used only two fitting parameters h_m and S.

Acknowledgments: Research reported in this publication was supported by the National Institute of General Medical Sciences of the National Institutes of Health under Award Number P30 GM110761. We are grateful to R.J. Collier of Harvard Medical School for plenteous advice and for the use of his lab equipment for cellular work. We are grateful to M.D.A. Myers for editorial assistance.

Author Contributions: A.S.L. conceived and designed the experiments and wrote the paper; M.V.-U. designed the experiments and wrote the paper; M.V.R. designed and performed translocation experiments and prepared mutants for all experiments; C.G. performed fluorescence experiments; O.S. performed protein synthesis inhibition assay.

Conflicts of Interest: The authors declare no conflict of interest. The founding sponsors had no role in the design of the study; in the collection, analyses, or interpretation of data; in the writing of the manuscript, and in the decision to publish the results.

References

1. Murphy, J.R. Mechanism of diphtheria toxin catalytic domain delivery to the eukaryotic cell cytosol and the cellular factors that directly participate in the process. *Toxins* **2011**, *3*, 294–308. [CrossRef] [PubMed]
2. Simon, N.C.; Aktories, K.; Barbieri, J.T. Novel bacterial ADP-ribosylating toxins: Structure and function. *Nat. Rev. Microbiol.* **2014**, *12*, 599–611. [CrossRef] [PubMed]
3. Zhang, Y.; Schulte, W.; Pink, D.; Phipps, K.; Zijlstra, A.; Lewis, J.D.; Waisman, D.M. Sensitivity of cancer cells to truncated diphtheria toxin. *PLoS ONE* **2010**, *5*, e10498. [CrossRef] [PubMed]
4. Wild, R.; Yokoyama, Y.; Dings, R.P.; Ramakrishnan, S. VEGF-DT385 toxin conjugate inhibits mammary adenocarcinoma development in a transgenic mouse model of spontaneous tumorigenesis. *Breast Cancer Res. Treat.* **2004**, *85*, 161–171. [CrossRef] [PubMed]
5. Urieto, J.O.; Liu, T.; Black, J.H.; Cohen, K.A.; Hall, P.D.; Willingham, M.C.; Pennell, L.K.; Hogge, D.E.; Kreitman, R.J.; Frankel, A.E. Expression and purification of the recombinant diphtheria fusion toxin DT388IL3 for phase i clinical trials. *Protein Expr. Purif.* **2004**, *33*, 123–133. [CrossRef] [PubMed]
6. Turturro, F. Denileukin diftitox: A biotherapeutic paradigm shift in the treatment of lymphoid-derived disorders. *Expert Rev. Anticancer Ther.* **2007**, *7*, 11–17. [CrossRef] [PubMed]
7. Ramakrishnan, S.; Olson, T.A.; Bautch, V.L.; Mohanraj, D. Vascular endothelial growth factor-toxin conjugate specifically inhibits KDR/flk-1-positive endothelial cell proliferation in vitro and angiogenesis in vivo. *Cancer Res.* **1996**, *56*, 1324–1330. [PubMed]
8. Ramage, J.G.; Vallera, D.A.; Black, J.H.; Aplan, P.D.; Kees, U.R.; Frankel, A.E. The diphtheria toxin/urokinase fusion protein (DTAT) is selectively toxic to CD87 expressing leukemic cells. *Leuk. Res.* **2003**, *27*, 79–84. [CrossRef]

9. Murphy, J.R.; Bishai, W.; Borowski, M.; Miyanohara, A.; Boyd, J.; Nagle, S. Genetic construction, expression, and melanoma-selective cytotoxicity of a diphtheria toxin-related alpha-melanocyte-stimulating hormone fusion protein. *Proc. Natl. Acad. Sci. USA* **1986**, *83*, 8258–8262. [CrossRef] [PubMed]

10. Kreitman, R.J. Immunotoxins for targeted cancer therapy. *AAPS J.* **2006**, *8*, E532–E551. [CrossRef] [PubMed]

11. Hogge, D.E.; Yalcintepe, L.; Wong, S.H.; Gerhard, B.; Frankel, A.E. Variant diphtheria toxin-interleukin-3 fusion proteins with increased receptor affinity have enhanced cytotoxicity against acute myeloid leukemia progenitors. *Clin. Cancer Res.* **2006**, *12*, 1284–1291. [CrossRef] [PubMed]

12. Hall, P.D.; Willingham, M.C.; Kreitman, R.J.; Frankel, A.E. DT$_{388}$-GM-CSF, a novel fusion toxin consisting of a truncated diphtheria toxin fused to human granulocyte-macrophage colony-stimulating factor, prolongs host survival in a scid mouse model of acute myeloid leukemia. *Leukemia* **1999**, *13*, 629–633. [CrossRef] [PubMed]

13. Feuring-Buske, M.; Frankel, A.; Gerhard, B.; Hogge, D. Variable cytotoxicity of diphtheria toxin 388-granulocyte-macrophage colony-stimulating factor fusion protein for acute myelogenous leukemia stem cells. *Exp. Hematol.* **2000**, *28*, 1390–1400. [CrossRef]

14. Duvic, M.; Talpur, R. Optimizing denileukin diftitox (ontak) therapy. *Future Oncol.* **2008**, *4*, 457–469. [CrossRef] [PubMed]

15. Cohen, K.A.; Liu, T.F.; Cline, J.M.; Wagner, J.D.; Hall, P.D.; Frankel, A.E. Toxicology and pharmacokinetics of DT388IL3, a fusion toxin consisting of a truncated diphtheria toxin (DT388) linked to human interleukin 3 (IL3), in cynomolgus monkeys. *Leuk. Lymphoma* **2004**, *45*, 1647–1656. [CrossRef] [PubMed]

16. Black, J.H.; McCubrey, J.A.; Willingham, M.C.; Ramage, J.; Hogge, D.E.; Frankel, A.E. Diphtheria toxin-interleukin-3 fusion protein (DT(388)IL3) prolongs disease-free survival of leukemic immunocompromised mice. *Leukemia* **2003**, *17*, 155–159. [CrossRef] [PubMed]

17. Reshetnyak, Y.K.; Andreev, O.A.; Lehnert, U.; Engelman, D.M. Translocation of molecules into cells by pH-dependent insertion of a transmembrane helix. *Proc. Natl. Acad. Sci. USA* **2006**, *103*, 6460–6465. [CrossRef] [PubMed]

18. Reshetnyak, Y.K.; Segala, M.; Andreev, O.A.; Engelman, D.M. A monomeric membrane peptide that lives in three worlds: In solution, attached to, and inserted across lipid bilayers. *Biophys. J.* **2007**, *93*, 2363–2372. [CrossRef] [PubMed]

19. Andreev, O.A.; Engelman, D.M.; Reshetnyak, Y.K. Targeting acidic diseased tissue: New technology based on use of the pH (Low) Insertion Peptide (pHLIP). *Chim. Oggi* **2009**, *27*, 34–37. [PubMed]

20. Segala, J.; Engelman, D.M.; Reshetnyak, Y.K.; Andreev, O.A. Accurate analysis of tumor margins using a fluorescent pH Low Insertion Peptide (pHLIP). *Int. J. Mol. Sci.* **2009**, *10*, 3478–3487. [CrossRef] [PubMed]

21. Vavere, A.L.; Biddlecombe, G.B.; Spees, W.M.; Garbow, J.R.; Wijesinghe, D.; Andreev, O.A.; Engelman, D.M.; Reshetnyak, Y.K.; Lewis, J.S. A novel technology for the imaging of acidic prostate tumors by positron emission tomography. *Cancer Res.* **2009**, *69*, 4510–4516. [CrossRef] [PubMed]

22. Reshetnyak, Y.K.; Yao, L.; Zheng, S.; Kuznetsov, S.; Engelman, D.M.; Andreev, O.A. Measuring tumor aggressiveness and targeting metastatic lesions with fluorescent pHLIP. *Mol. Imaging Biol.* **2011**, *13*, 1146–1156. [CrossRef] [PubMed]

23. Andreev, O.A.; Engelman, D.M.; Reshetnyak, Y.K. pH-sensitive membrane peptides (pHLIPs) as a novel class of delivery agents. *Mol. Membr. Biol.* **2010**, *27*, 341–352. [CrossRef] [PubMed]

24. Sosunov, E.A.; Anyukhovsky, E.P.; Sosunov, A.A.; Moshnikova, A.; Wijesinghe, D.; Engelman, D.M.; Reshetnyak, Y.K.; Andreev, O.A. pH (Low) Insertion Peptide (pHLIP) targets ischemic myocardium. *Proc. Natl. Acad. Sci. USA* **2013**, *110*, 82–86. [CrossRef] [PubMed]

25. Weerakkody, D.; Moshnikova, A.; Thakur, M.S.; Moshnikova, V.; Daniels, J.; Engelman, D.M.; Andreev, O.A.; Reshetnyak, Y.K. Family of pH (Low) Insertion Peptides for tumor targeting. *Proc. Natl. Acad. Sci. USA* **2013**, *110*, 5834–5839. [CrossRef] [PubMed]

26. Andreev, O.A.; Engelman, D.M.; Reshetnyak, Y.K. Targeting diseased tissues by pHLIP insertion at low cell surface pH. *Front. Physiol.* **2014**, *5*, 97. [CrossRef] [PubMed]

27. Ren, J.; Kachel, K.; Kim, H.; Malenbaum, S.E.; Collier, R.J.; London, E. Interaction of diphtheria toxin T domain with molten globule-like proteins and its implications for translocation. *Science* **1999**, *284*, 955–957. [CrossRef] [PubMed]

28. Ladokhin, A.S. pH-triggered conformational switching along the membrane insertion pathway of the diphtheria toxin T-domain. *Toxins* **2013**, *5*, 1362–1380. [CrossRef] [PubMed]

29. Ladokhin, A.S.; Legmann, R.; Collier, R.J.; White, S.H. Reversible refolding of the diphtheria toxin *T*-domain on lipid membranes. *Biochemistry* **2004**, *43*, 7451–7458. [CrossRef] [PubMed]

30. Kyrychenko, A.; Posokhov, Y.O.; Rodnin, M.V.; Ladokhin, A.S. Kinetic intermediate reveals staggered pH-dependent transitions along the membrane insertion pathway of the diphtheria toxin *T*-domain. *Biochemistry* **2009**, *48*, 7584–7594. [CrossRef] [PubMed]

31. Vargas-Uribe, M.; Rodnin, M.V.; Ojemalm, K.; Holgado, A.; Kyrychenko, A.; Nilsson, I.; Posokhov, Y.O.; Makhatadze, G.; von Heijne, G.; Ladokhin, A.S. Thermodynamics of membrane insertion and refolding of the diphtheria toxin *T*-domain. *J. Membr. Biol.* **2015**, *248*, 383–394. [CrossRef] [PubMed]

32. Rodnin, M.V.; Kyrychenko, A.; Kienker, P.; Sharma, O.; Posokhov, Y.O.; Collier, R.J.; Finkelstein, A.; Ladokhin, A.S. Conformational switching of the diphtheria toxin T domain. *J. Mol. Biol.* **2010**, *402*, 1–7. [CrossRef] [PubMed]

33. Kurnikov, I.V.; Kyrychenko, A.; Flores-Canales, J.C.; Rodnin, M.V.; Simakov, N.; Vargas-Uribe, M.; Posokhov, Y.O.; Kurnikova, M.; Ladokhin, A.S. Ph-triggered conformational switching of the diphtheria toxin *T*-domain: The roles of *N*-terminal histidines. *J. Mol. Biol.* **2013**, *425*, 2752–2764. [CrossRef] [PubMed]

34. Rodnin, M.V.; Li, J.; Gross, M.L.; Ladokhin, A.S. The pH-dependent trigger in diphtheria toxin *T* domain comes with a safety latch. *Biophys. J.* **2016**, *111*, 1946–1953. [CrossRef] [PubMed]

35. Perier, A.; Chassaing, A.; Raffestin, S.; Pichard, S.; Masella, M.; Menez, A.; Forge, V.; Chenal, A.; Gillet, D. Concerted protonation of key histidines triggers membrane interaction of the diphtheria toxin *T* domain. *J. Biol. Chem.* **2007**, *282*, 24239–24245. [CrossRef] [PubMed]

36. Ghatak, C.; Rodnin, M.V.; Vargas-Uribe, M.; McCluskey, A.J.; Flores-Canales, J.C.; Kurnikova, M.; Ladokhin, A.S. Role of acidic residues in helices TH8–TH9 in membrane interactions of the diphtheria toxin *T* domain. *Toxins* **2015**, *7*, 1303–1323. [CrossRef] [PubMed]

37. Rodnin, M.V.; Kyrychenko, A.; Kienker, P.; Sharma, O.; Vargas-Uribe, M.; Collier, R.J.; Finkelstein, A.; Ladokhin, A.S. Replacement of *C*-terminal histidines uncouples membrane insertion and translocation in diphtheria toxin *T*-domain. *Biophys. J.* **2011**, *101*, L41–L43. [CrossRef] [PubMed]

38. Vargas-Uribe, M.; Rodnin, M.V.; Kienker, P.; Finkelstein, A.; Ladokhin, A.S. Crucial role of H322 in folding of the diphtheria toxin *T*-domain into the Open-Channel State. *Biochemistry* **2013**, *52*, 3457–3463. [CrossRef] [PubMed]

39. Kienker, P.K.; Wu, Z.; Finkelstein, A. Mapping the membrane topography of the TH6–TH7 segment of the diphtheria toxin *T*-domain channel. *J. Gen. Physiol.* **2015**, *145*, 107–125. [CrossRef] [PubMed]

40. Kienker, P.K.; Wu, Z.; Finkelstein, A. Topography of the TH5 segment in the diphtheria toxin *T*-domain channel. *J. Membr. Biol.* **2016**, *249*, 181–196. [CrossRef] [PubMed]

41. Donovan, J.J.; Simon, M.I.; Draper, R.K.; Montal, M. Diphtheria toxin forms transmembrane channels in planar lipid bilayers. *Proc. Natl. Acad. Sci. USA* **1981**, *78*, 172–176. [CrossRef] [PubMed]

42. Bennett, M.J.; Choe, S.; Eisenberg, D. Refined structure of dimeric diphtheria toxin at 2.0 Å resolution. *Protein Sci.* **1994**, *3*, 1444–1463. [CrossRef] [PubMed]

43. Senzel, L.; Gordon, M.; Blaustein, R.O.; Oh, K.J.; Collier, R.J.; Finkelstein, A. Topography of diphtheria toxin's T domain in the open channel state. *J. Gen. Physiol.* **2000**, *115*, 421–434. [CrossRef] [PubMed]

44. Oh, K.J.; Zhan, H.; Cui, C.; Hideg, K.; Collier, R.J.; Hubbell, W.L. Organization of diphtheria toxin T domain in bilayers: A site-directed spin labeling study. *Science* **1996**, *273*, 810–812. [CrossRef] [PubMed]

45. Mindell, J.A.; Silverman, J.A.; Collier, R.J.; Finkelstein, A. Structure-function relationships in diphtheria toxin channels: III. Residues which affect the *cis* pH dependence of channel conductance. *J. Membr. Biol.* **1994**, *137*, 45–57. [CrossRef] [PubMed]

46. Silverman, J.A.; Mindell, J.A.; Finkelstein, A.; Shen, W.H.; Collier, R.J. Mutational analysis of the helical hairpin region of diphtheria toxin transmembrane domain. *J. Biol. Chem.* **1994**, *269*, 22524–22532. [PubMed]

47. Silverman, J.A.; Mindell, J.A.; Zhan, H.; Finkelstein, A.; Collier, R.J. Structure-function relationships in diphtheria toxin channels: I. Determining a minimal channel-forming domain. *J. Membr. Biol.* **1994**, *137*, 17–28. [CrossRef] [PubMed]

48. Kaul, P.; Silverman, J.; Shen, W.H.; Blanke, S.R.; Huynh, P.D.; Finkelstein, A.; Collier, R.J. Roles of Glu 349 and Asp 352 in membrane insertion and translocation by diphtheria toxin. *Protein Sci.* **1996**, *5*, 687–892. [CrossRef] [PubMed]

49. Rodnin, M.V.; Ladokhin, A.S. Membrane translocation assay based on proteolytic cleavage: Application to diphtheria toxin T domain. *Biochim. Biophys. Acta* **2015**, *1848*, 35–40. [CrossRef] [PubMed]

50. Falnes, P.O.; Madshus, I.H.; Sandvig, K.; Olsnes, S. Replacement of negative by positive charges in the presumed membrane-inserted part of diphtheria toxin B fragment. Effect on membrane translocation and on formation of cation channels. *J. Biol. Chem.* **1992**, *267*, 12284–12290. [PubMed]

51. London, E.; Ladokhin, A.S. Measuring the depth of amino acid residues in membrane-inserted peptides by fluorescence quenching. *Curr. Top. Membr.* **2002**, *52*, 89–115.

52. Ladokhin, A.S. Measuring membrane penetration with depth-dependent fluorescence quenching: Distribution analysis is coming of age. *Biochim. Biophys. Acta* **2014**, *1838*, 2289–2295. [CrossRef] [PubMed]

53. Kyrychenko, A.; Posokhov, Y.O.; Vargas-Uribe, M.; Ghatak, C.; Rodnin, M.V.; Ladokhin, A.S. Fluorescence applications for structural and thermodynamic studies of membrane protein insertion. In *Reviews in Fluorescence 2016*; Geddes, C.D., Ed.; Springer: Cham, Switzerland, 2017; pp. 243–274.

54. McIntosh, T.J.; Holloway, P.W. Determination of the depth of bromine atoms in bilayers formed from bromolipid probes. *Biochemistry* **1987**, *26*, 1783–1788. [CrossRef] [PubMed]

55. Ladokhin, A.S. Analysis of protein and peptide penetration into membranes by depth-dependent fluorescence quenching: Theoretical considerations. *Biophys. J.* **1999**, *76*, 946–955. [CrossRef]

56. Ladokhin, A.S. Evaluation of lipid exposure of tryptophan residues in membrane peptides and proteins. *Anal. Biochem.* **1999**, *276*, 65–71. [CrossRef] [PubMed]

57. Gnanasambandam, R.; Ghatak, C.; Yasmann, A.; Nishizawa, K.; Sachs, F.; Ladokhin, A.S.; Sukharev, S.I.; Suchyna, T.M. GsMTx4: Mechanism of inhibiting mechanosensitive ion channels. *Biophys. J.* **2017**, *112*, 31–45. [CrossRef] [PubMed]

58. Blanke, S.R.; Milne, J.C.; Benson, E.L.; Collier, R.J. Fused polycationic peptide mediates delivery of diphtheria toxin A chain to the cytosol in the presence of anthrax protective antigen. *Proc. Natl. Acad. Sci. USA* **1996**, *93*, 8437–8442. [CrossRef] [PubMed]

59. Hope, M.J.; Bally, M.B.; Mayer, L.D.; Janoff, A.S.; Cullis, P.R. Generation of multilamellar and unilamellar phospholipid vesicles. *Chem. Phys. Lipids* **1986**, *40*, 89–107. [CrossRef]

60. Mayer, L.D.; Hope, M.J.; Cullis, P.R. Vesicles of variable sizes produced by a rapid extrusion procedure. *Biochim. Biophys. Acta* **1986**, *858*, 161–168. [CrossRef]

61. Barbieri, J.T.; Collier, R.J. Expression of a mutant, full-length form of diphtheria toxin in *Escherichia coli.*. *Infect Immun.* **1987**, *55*, 1647–1651. [PubMed]

62. Zhan, H.; Elliott, J.L.; Shen, W.H.; Huynh, P.D.; Finkelstein, A.; Collier, R.J. Effects of mutations in proline 345 on insertion of diptheria toxin into model membranes. *J. Membr. Biol.* **1999**, *167*, 173–181. [CrossRef] [PubMed]

63. Ladokhin, A.S.; Jayasinghe, S.; White, S.H. How to measure and analyze tryptophan fluorescence in membranes properly, and why bother? *Anal. Biochem.* **2000**, *285*, 235–245. [CrossRef] [PubMed]

toxins

MDPI

Perspective

The Unexpected Tuners: Are LncRNAs Regulating Host Translation during Infections?

Primoz Knap [1], Toma Tebaldi [2], Francesca Di Leva [3], Marta Biagioli [3], Mauro Dalla Serra [1] and Gabriella Viero [1,*]

[1] Institute of Biophysics, CNR Unit at Trento, Via Sommarive 18, Povo Trento 38123, Italy; primoz.knap@gmail.com (P.K.); mauro.dallaserra@cnr.it (M.D.S.)
[2] Yale Cancer Center, Yale University School of Medicine, New Haven, CT 06520, USA; toma.tebaldi@yale.edu
[3] Centre for Integrative Biology, University of Trento, Via Sommarive 9, Povo Trento 38123, Italy; francesca.dileva@unitn.it (F.D.L.); marta.biagioli@unitn.it (M.B.)
* Correspondence: gabriella.viero@cnr.it; Tel.: +39-0461-314033

Academic Editor: Alexey S. Ladokhin
Received: 10 October 2017; Accepted: 31 October 2017; Published: 3 November 2017

Abstract: Pathogenic bacteria produce powerful virulent factors, such as pore-forming toxins, that promote their survival and cause serious damage to the host. Host cells reply to membrane stresses and ionic imbalance by modifying gene expression at the epigenetic, transcriptional and translational level, to recover from the toxin attack. The fact that the majority of the human transcriptome encodes for non-coding RNAs (ncRNAs) raises the question: do host cells deploy non-coding transcripts to rapidly control the most energy-consuming process in cells—i.e., host translation—to counteract the infection? Here, we discuss the intriguing possibility that membrane-damaging toxins induce, in the host, the expression of toxin-specific long non-coding RNAs (lncRNAs), which act as sponges for other molecules, encoding small peptides or binding target mRNAs to depress their translation efficiency. Unravelling the function of host-produced lncRNAs upon bacterial infection or membrane damage requires an improved understanding of host lncRNA expression patterns, their association with polysomes and their function during this stress. This field of investigation holds a unique opportunity to reveal unpredicted scenarios and novel approaches to counteract antibiotic-resistant infections.

Keywords: host–pathogen interaction; bacterial toxins; pore-forming toxins (PFTs); long non-coding RNAs (lncRNAs); translation; translational control; ribosome profiling; polysome profiling

1. Membrane Damaging Toxins, Osmotic Imbalance and Translation

According to the World Health Organization (WHO), drug-resistant pathogenic bacteria are estimated to cause 25,000 deaths every year in the European Union alone. These bacteria also lead to high medical costs, prolonged hospital stays and increased mortality [1]. Bacterial pathogens possess a plethora of strategies to subvert host defenses, by secreting biological macromolecules, such as toxins [2,3], which promote bacterial survival within the host environment, for example, by escaping recognition from the immune response. Infections mediated by pathogens often impact the protein synthesis efficiency of the host cell, limiting the production of proteins involved in cellular recovery, like cytokines [4]. Protein synthesis is, in fact, the most energy consuming cellular process, justifying why cells have evolved finely tuned translational control mechanisms to conserve energy and respond quickly to stimuli, if needed [5]. Thus, it is reasonable to presume that translation regulation should be tightly modulated upon bacterial infection through, as yet, poorly understood mechanisms.

One of the most ancient forms of attack exerted by bacterial virulence factors is the formation of proteinaceous pores that cross plasmatic or intracellular membranes [3,6]. These proteins, called

Pore-Forming Toxins (PFTs), account for about 25–30% of all bacterial toxins [7]. The intimate relationship between different PFTs and host cell membranes is based on an amazingly large variety of highly specific interactions between toxins and various types of host receptors: sugars, membrane lipids or proteins. While different PFTs use different binding strategies, they all share a common multi-step mechanism of action, for pore formation: (i) release of water-soluble monomers, (ii) binding of monomers to the target membrane, (iii) oligomerization in a non-lytic pre-pore, (iv) insertion of the pore-forming protein portion into the lipid bilayer and opening of nanosized aqueous pores in the host membrane [3,8]. At high toxin doses, this intimate inter-species interaction leads to a massive number of pores, followed by an ionic imbalance [9,10] and indirect or direct membrane damage [8,11]. Cells reply to the osmotic stress by deploying sophisticated mechanisms that counteract the damaging effects of toxins [9,12]. If the activation of host survival or membrane repair mechanisms [10] does not succeed in opposing the stress, cells die, via apoptosis, necrosis or membrane damage. Activation of autophagy and necroptosis have been described as responses to many PFTs, such as aerolysin, vibrio cholerae cytolysin (VCC), *S. aureus* cytolysins [3] and listeriolysin O (LLO) [13]. Even at sub-lytic doses, the binding of toxin monomers or the insertion of a few pores into membranes are still able to provoke extremely diverse cellular responses [11,14]. In fact, the local perturbation of the lipid bilayer upon toxin binding can impact the physiology of the host membrane, by rewiring the physico-chemical organization of the lipid bilayer and altering the proper functionality of host membrane proteins involved in intracellular signaling [15,16].

The proteinaceous pores formed in the host membrane have a wide variety of ionic selectivity and distribution of lumen diameters, ranging from few to tens of nanometers [17]. In any case the pore induces a re-equilibration of ion concentrations across the plasma membrane, resulting in calcium influx and potassium efflux. By a still unclear mechanism, cells are able to detect decreases in the cytosolic potassium concentration, caused by changes in membrane permeability [18]. Calcium is a potent secondary messenger in cells and its ionic imbalance triggers the activation of various signaling cascades to repair the damaged membrane and restore homeostasis. Calcium release from intracellular stores was shown to induce Endoplasmic Reticulum (ER) stress, activating the Unfolded Protein Response (UPR), Ca^{2+} dependent proteases, and Ca^{2+} dependent membrane repair strategies [19]. In addition, the activation of several defense mechanisms, such as MAPK/p38/ERK/JNK, AKT/mTORC pathways [3,18,20] and the inflammasome complex, have been observed [18].

All these events act in concert to control protein synthesis. Potassium efflux induces a transient stop in protein synthesis upon PFT treatment [14,18], a somehow expected outcome since translation can be controlled directly [21] or indirectly through ion fluxes [22]. Moreover, the abovementioned activation of MAPK/p38/ERK/JNK and AKT/mTORC controls the functionality of general translational factors, i.e., eIF4E, eIF2α and eEF2 [5]. Similarly, the crosstalk between potassium efflux and calcium influx can activate the PERK signaling pathway through the UPR, a sensor of ER stress. PERK controls translation via phosphorylation of eIF2α, thereby globally suppressing translation initiation [23]. Overall, the equilibrium between activation and inactivation of translation factors allows cells to enter a low-energy consumption state, in parallel to a rewiring of protein synthesis. Such expedients can facilitate cell survival until recovery of membrane integrity, pointing towards translation as a major hub in promoting cell endurance upon infection and osmotic stress.

Despite this evidence, very few studies have explored the global landscape of changes at the translational (Table 1) or transcriptional [7] levels, occurring as a host response to virulent attacks. Indeed, most of the available studies have focused on transcriptional variations induced by defined immune-stimulatory ligands, such as lipopolysaccharide, with a very recent exception where the host translation response to pathogen infection was monitored by ribosome profiling [24]. Given these still sparse observations, a clear gap of knowledge exists on the precise involvement of translational control in tuning host protein synthesis after exposure to pathogens. This fact preludes a new and interesting field of investigation.

Table 1. Genome-wide translatome/protein synthesis analyses of host response to virulent factors.

Method	System	Reference
Ribosome profiling	macrophages infected with the intracellular bacterial pathogen *Legionella pneumophila*	[24]
Ribosome profiling	macrophages treated with LPS	[25]
Sucrose gradient ultracentrifugation followed by microarray analysis	SH-SY5Y cells treated with lytic and sub-lytic doses of α-haemolysin	[14]
Pulsed SILAC proteomics	dendritic cells treated with LPS	[26]
Sucrose gradient ultracentrifugation followed by PCR array analysis	RAW 264.7 murine macrophages treated with ribotoxic mycotoxin DON	[27]
Sucrose gradient ultracentrifugation followed by microarray analysis	human monocyte-derived dendritic cells treated with LPS	[28]
Sucrose gradient ultracentrifugation followed by microarray analysis	macrophage-like J774.1 cells treated with LPS	[29]

2. Host Long Non-Coding RNAs (LncRNAs): An Overlooked Toolkit for Controlling Gene Expression in Host–Pathogen Interaction Studies

Non-coding RNAs (ncRNAs) are very good candidates for the specific and tight regulation of protein synthesis in cells experiencing stresses, such as pore formation and ionic imbalance. Among ncRNAs, long non-coding RNAs (lncRNAs) represent a long-time neglected class of molecules, found in animals and plants. What is striking is that in humans, the number of genes encoding for lncRNAs almost matches the number of protein-coding genes [30]. Importantly, the Encyclopedia of DNA Elements (ENCODE) project, as well as the RIKEN Functional Annotation of the Mammalian Genome (FANTOM 5) consortium, proposed a biochemical function for most lncRNAs. Even if the scientific community is far from being concordant on this matter, with many scientists arguing that the term "functional" is misleading, it is possible that the production of these RNAs represents an ideal playground for evolving new mechanisms to control gene expression across all levels, from transcription to translation.

LncRNAs are a sub-group of non-coding RNAs, loosely defined as transcripts that are longer than 200 nt with no apparent protein coding potential. They can be classified according to two criteria: their genomic position and their mechanism of action or function (Table 2). A significant fraction of lncRNAs appears to be 5′-capped and polyadenylated [31], and presents a similar chromatin arrangement to their actively-transcribed, protein-coding counterparts [32]. However, they do share some common characteristics that distinguish them from *bona fide* protein coding mRNAs (Table 3).

Table 2. Classification of lncRNAs according to genomic position or mechanism function.

Genomic Position			Mechanism or Function		
Name	*Description*	*Reference*	*Name*	*Description*	*Reference*
Intergenic lncRNAs (lincRNAs)	do not overlap with any part of a protein coding gene and are at least 1 kb distant from it	[33]	Competing endogenous RNAs (ceRNAs)	also called miRNA "sponges", which participate in a microRNA-dependent crosstalk. These lncRNAs share miRNA response elements (MREs) with some mRNAs, thereby sequestering miRNAs	[34]

Table 2. *Cont.*

Genomic Position			Mechanism or Function		
Name	*Description*	*Reference*	*Name*	*Description*	*Reference*
Trans-Natural Antisense Transcripts (NATs)	antisense lncRNAs acting on mRNAs and complementary to transcripts from remote loci.	[35]	Protein "sponges"	bind regulatory proteins, disabling them from interacting with their potential targets	[36]
Cis-Natural Antisense Transcripts (NATs)	antisense lncRNAs acting on mRNAs. These lncRNAs *are transcribed* from the same genomic region as their complementary sense transcript	[35]	Scaffolding lncRNAs	act as a scaffold for multiple chromatin remodelling complexes	[37]
Sense-overlapping or transcribed pseudogene lncRNAs	are considered transcript variants of protein coding mRNAs, and overlap with a protein coding gene on the same DNA strand	[38]	SINEUPs	antisense lncRNAs that stimulate cap-independent translation of target sense mRNAs through the activity of an embedded repetitive element	[39,40]
Intronic lncRNAs	located in the introns of protein coding genes without overlapping with their exons	[41]	Stress-induced lncRNAs (silncRNAs)	Induced upon cell stress, permit a faster recovery of the cell cycle delay caused by stress	[42]
			Modulators of Post Translational Modifications	Act on post-translational modifications of proteins, such as ubiquitination and phosphorylation	[43]

Table 3. Characteristics of lncRNAs.

Features	Reference
Lack of a single long open reading frame (ORF) > 300 nt	[44,45]
Low expression levels, compared to mRNAs	[46,47]
Longer but fewer exons than protein-coding genes, with a bias toward two-exons transcripts	[48]
Exons with a significantly lower GC content, compared to protein-coding RNAs	[44]
Paucity or absence of introns	[44]
Enrichments in nucleus	[49]
High degree of tissue specificity	[46,48]
Co-expression with neighboring genes	[46]
Low evolutionary conservation of primary sequence	[50]

Interestingly, a growing amount of evidence supports the involvement of lncRNAs in regulating post-transcriptional processes and translation [39]. Surprisingly, several lncRNAs have been found to associate with ribosomes [51] and polysomes containing one, two or three ribosomes [52,53]. As to what function they may perform on translation is still a matter of debate. Ribosome profiling experiments have demonstrated that several lncRNAs are in fact engaged by ribosomes as mRNAs [51], raising questions about their classification as non-coding. In accordance with this finding, some lncRNAs were in fact shown to produce short peptides [54] with still unknown functions. Alternative hypotheses to short-peptide production, propose that lncRNAs can rather serve as scaffolds or regulatory platforms, facilitating the recruitment of mRNAs on polysomes. New natural antisense lncRNA classes, that hybridize head-to-head to protein-coding genes, have been described as stimulating cap-independent and cap-dependent translation of target sense mRNAs [55]. These antisense lncRNAs are in fact able to specifically bind to the corresponding sense transcript and, by a still debated mechanism, function as 'ribosome recruiters'. The "ribosome recruitment" activity of these lncRNAs, implicated in cap-independent translation, resides on embedded repetitive elements (SINEB2 in mouse and FRAM/MIRb in human) [39,40], likely inducing a peculiar RNA structural organization that acts as a scaffold for ribosome engagement to protein-coding transcripts. Importantly, this activity is completely independent of transcriptional effects in the sense transcript [39,40].

It is fair to say that when it comes to the involvement of ncRNAs in host–pathogen interaction studies, the class of lncRNAs has taken the back seat. Most studies on host–pathogen crosstalk are focused on the role played by small ncRNAs, specifically miRNAs [56–59]. Studies addressing specific lncRNAs are mainly restricted to their involvement in viral infection [43,56–59]. Indeed, studies considering the role of host lncRNAs transcription during bacterial infection are currently limited to only a handful of examples, nicely reviewed very recently [60]. These pioneering studies shed light on lncRNAs possibly playing an important role in the cell's response to bacterial infection or in the induction of inflammation, through Toll-like receptor ligands [61].

3. Are LncRNAs Overlooked Translation Regulators in Host–Pathogen Crosstalk?

To our knowledge, no information has been collected yet concerning the involvement of lncRNAs in modulating host translation upon either bacterial infection or treatment with virulent factors as PFTs. Indeed, we illustrated that translation is a major hotspot amid the host–pathogen fight for survival. Hence, the impact of PFTs in tuning host protein synthesis efficiency to limit the production of proteins, by triggering the expression and direct recruitment of lncRNAs on ribosomes or polysomes, is likely more than a simple hypothesis.

Interestingly, combined transcriptomics and proteomics have demonstrated that during hyperosmotic stress, yeast is able to adapt by deploying numerous lncRNAs. The transcriptional interplay between stress-activated protein kinases and the induction of a number of non-coding transcripts [42,62], in turn, regulates the transcription of mRNAs coding for downstream factors of the MAPK pathway [62]. Despite the lack of clear mechanisms of action, these lncRNAs have the potential to induce a time-controlled depression of protein synthesis of their target transcripts, helping adaptation to hyperosmotic conditions [62]. Moreover, in higher eukaryotes, evidence accrued over the very last few years has revealed several examples of associations between lncRNA expression and regulation of the MAPK and AKT pathways in cancer [63–65] or of the PERK pathway and ER stress in viral infections [66]. Even if the cause and effect relationship between lncRNA expression and modulation of these well-known pathways is not yet clear, it is tempting to speculate that host cells could take advantage of this class of ncRNAs to finely tune translation and cope with the ionic imbalance triggered by PFT attack (Figure 1).

Given these observations, discovering the functions of other infection-induced lncRNAs and determining their mechanism of action will unquestionably expand our knowledge of the host–pathogen crosstalk. Ribosome profiling and polysome profiling experiments, performed in cells treated with pathogenic bacteria, could greatly improve our comprehension of the role of infection-induced lncRNAs in translation control. Comparing the host's response to invading bacterial strains, either expressing or lacking specific virulent factors, may give valuable insight into their role in the host–pathogen crosstalk, yielding important advances in understanding the interaction between organisms. Moreover, integration of in vivo and in vitro studies, using silencing and in vitro translation systems, can help to address the coding or non-coding functions of several lncRNAs, already found to be up- or down-regulated in cells, upon exposure to virulent factors. Therefore, further research on how cells use lncRNAs to cope with either bacterial infection or the damage caused by PFTs has a huge potential for unveiling, till now unforeseen, scenarios that might shed new light on host–pathogen crosstalk and reveal, as yet unpredicted, approaches to counteract antibiotic-resistant infections.

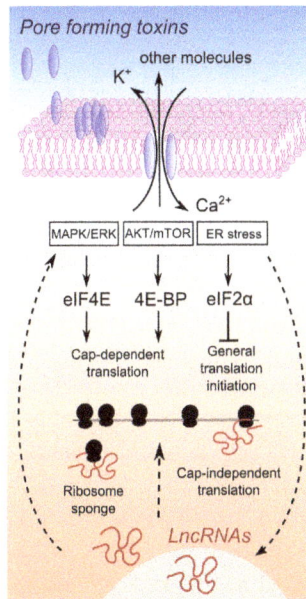

Figure 1. Hypothesis of interplay between lncRNA expression changes and the control of protein synthesis upon pore formation. Upon pore formation, efflux of potassium ions and influx of calcium ions are well known to occur, due to the activity of a large variety of PFTs. A simplified connection between ion imbalance and the activation of three major pathways is depicted (for a complete discussion please refer to the excellent review in [12]). These pathways control downstream target proteins, which are general factors of translation. Straight arrows connect processes related to the activation of pathways that control translation, proven to be involved in the response to ion imbalance triggered by pore-forming toxins or bacterial pathogens. In several cases, an association between lncRNA expression changes and regulation of these pathways has been demonstrated in cancer [63–65] or viral infections [66]. The cause and effect relationship of lncRNAs expression and the activation of pathways that control translation is at present not clear, as well as the mechanism of action behind such a connection. Therefore, we used dashed arrows to link lncRNA expression changes to pathways controlling translation, a connection that has been demonstrated for some lncRNAs but not with respect to bacterial infections, ion imbalance or pore formation by bacterial virulent factors.

Acknowledgments: This research was partially performed in the framework of "Grandi Progetti 2012" funded by Autonomous Province of Trento (PAT), Italy: "Axonomics: identifying the translational networks altered in motor neuron diseases—AXonomIX Project" (319266-S116/2013). P.K. research work has been supported by a fellowship in the memory of Gianfranco Menestrina.

Author Contributions: P.K. and G.V. conceived the review, P.K., T.T., F.D.L., M.B., M.D.S. and G.V. performed research in literature and wrote the paper; P.K. and G.V. prepared Tables 2 and 3, T.T. prepared Table 1.

Conflicts of Interest: The authors declare no conflict of interest.

References

1. World Health Organization. *Global Action Plan on Antimicrobial Resistance*; World Health Organization: Geneva, Switzerland, 2015.
2. Schiavo, G.; van der Goot, F.G. The bacterial toxin toolkit. *Nat. Rev. Mol. Cell Biol.* **2001**, *2*, 530–537. [CrossRef] [PubMed]
3. Dal Peraro, M.; van der Goot, F.G. Pore-forming toxins: Ancient, but never really out of fashion. *Nat. Rev. Microbiol.* **2016**, *14*, 77–92. [CrossRef] [PubMed]

4. Shrestha, N.; Bahnan, W.; Wiley, D.J.; Barber, G.; Fields, K.A.; Schesser, K. Eukaryotic initiation factor 2 (eIF2) signaling regulates proinflammatory cytokine expression and bacterial invasion. *J. Biol. Chem.* **2012**, *287*, 28738–28744. [CrossRef] [PubMed]

5. Roux, P.P.; Topisirovic, I. Regulation of mRNA translation by signaling pathways. *Cold Spring Harb. Perspect. Biol.* **2012**, *4*. [CrossRef] [PubMed]

6. Menestrina, G.; Dalla Serra, M.; Comai, M.; Coraiola, M.; Viero, G.; Werner, S.; Colin, D.A.; Monteil, H.; Prévost, G. Ion channels and bacterial infection: The case of beta-barrel pore-forming protein toxins of Staphylococcus aureus. *FEBS Lett.* **2003**, *552*, 54–60. [CrossRef]

7. Kao, C.-Y.; Los, F.C.O.; Huffman, D.L.; Wachi, S.; Kloft, N.; Husmann, M.; Karabrahimi, V.; Schwartz, J.-L.; Bellier, A.; Ha, C.; et al. Global functional analyses of cellular responses to pore-forming toxins. *PLoS Pathog.* **2011**, *7*, e1001314. [CrossRef] [PubMed]

8. Menestrina, G.; Dalla Serra, M.; Prévost, G. Mode of action of beta-barrel pore-forming toxins of the staphylococcal gamma-hemolysin family. *Toxicon* **2001**, *39*, 1661–1672. [CrossRef]

9. Kloft, N.; Busch, T.; Neukirch, C.; Weis, S.; Boukhallouk, F.; Bobkiewicz, W.; Cibis, I.; Bhakdi, S.; Husmann, M. Pore-forming toxins activate MAPK p38 by causing loss of cellular potassium. *Biochem. Biophys. Res. Commun.* **2009**, *385*, 503–506. [CrossRef] [PubMed]

10. Wolfmeier, H.; Schoenauer, R.; Atanassoff, A.P.; Neill, D.R.; Kadioglu, A.; Draeger, A.; Babiychuk, E.B. Ca^{2+}-dependent repair of pneumolysin pores: A new paradigm for host cellular defense against bacterial pore-forming toxins. *Biochim. Biophys. Acta Mol. Cell Res.* **2015**, *1853*, 2045–2054. [CrossRef] [PubMed]

11. Bischofberger, M.; Iacovache, I.; Gisou van der Goot, F. Pathogenic Pore-Forming Proteins: Function and Host Response. *Cell Host Microbe* **2012**, *12*, 266–275. [CrossRef] [PubMed]

12. Los, F.C.O.; Randis, T.M.; Aroian, R.V.; Ratner, A.J. Role of Pore-Forming Toxins in Bacterial Infectious Diseases. *Microbiol. Mol. Biol. Rev.* **2013**, *77*, 173–207. [CrossRef] [PubMed]

13. Meixenberger, K.; Pache, F.; Eitel, J.; Schmeck, B.; Hippenstiel, S.; Slevogt, H.; N'Guessan, P.; Witzenrath, M.; Netea, M.G.; Chakraborty, T.; et al. Listeria monocytogenes-Infected Human Peripheral Blood Mononuclear Cells Produce IL-1, Depending on Listeriolysin O and NLRP3. *J. Immunol.* **2010**, *184*, 922–930. [CrossRef] [PubMed]

14. Clamer, M.; Tebaldi, T.; Marchioretto, M.; Bernabo, P.; Bertini, E.; Guella, G.; Serra, M.D.; Quattrone, A.; Viero, G. Global translation variations in host cells upon attack of lytic and sublytic Staphylococcus aureus-haemolysin. *Biochem. J.* **2015**, *472*, 83–95. [CrossRef] [PubMed]

15. Chong, P.A.; Forman-Kay, J.D. Liquid-liquid phase separation in cellular signaling systems. *Curr. Opin. Struct. Biol.* **2016**, *41*, 180–186. [CrossRef] [PubMed]

16. Li, P.; Banjade, S.; Cheng, H.-C.; Kim, S.; Chen, B.; Guo, L.; Llaguno, M.; Hollingsworth, J.V.; King, D.S.; Banani, S.F.; et al. Phase transitions in the assembly of multivalent signalling proteins. *Nature* **2012**, *483*, 336–340. [CrossRef] [PubMed]

17. Gilbert, R.J.C.; Serra, M.D.; Froelich, C.J.; Wallace, M.I.; Anderluh, G. Membrane pore formation at protein-lipid interfaces. *Trends Biochem. Sci.* **2014**, *39*, 510–516. [CrossRef] [PubMed]

18. Gonzalez, M.R.; Bischofberger, M.; Frêche, B.; Ho, S.; Parton, R.G.; van der Goot, F.G. Pore-forming toxins induce multiple cellular responses promoting survival. *Cell. Microbiol.* **2011**, *13*, 1026–1043. [CrossRef] [PubMed]

19. Cooper, S.T.; McNeil, P.L. Membrane Repair: Mechanisms and Pathophysiology. *Physiol. Rev.* **2015**, *95*, 1205–1240. [CrossRef] [PubMed]

20. Chakravorty, A.; Awad, M.M.; Cheung, J.K.; Hiscox, T.J.; Lyras, D.; Rood, J.I. The pore-forming α-toxin from clostridium septicum activates the MAPK pathway in a Ras-c-Raf-dependent and independent manner. *Toxins* **2015**, *7*, 516–534. [CrossRef] [PubMed]

21. Iordanov, M.S.; Magun, B.E. Loss of cellular K+ mimics ribotoxic stress. Inhibition of protein synthesis and activation of the stress kinases SEK1/MKK4, stress-activated protein kinase/c-Jun NH2-terminal kinase 1, and p38/HOG1 by palytoxin. *J. Biol. Chem.* **1998**, *273*, 3528–3534. [CrossRef] [PubMed]

22. Tawk, M.Y.; Zimmermann-Meisse, G.; Bossu, J.L.; Potrich, C.; Bourcier, T.; Dalla Serra, M.; Poulain, B.; Prévost, G.; Jover, E. Internalization of staphylococcal leukotoxins that bind and divert the C5a receptor is required for intracellular Ca^{2+} mobilization by human neutrophils. *Cell. Microbiol.* **2015**, *17*, 1241–1257. [CrossRef] [PubMed]

23. Sonenberg, N.; Hinnebusch, A.G. Regulation of translation initiation in eukaryotes: Mechanisms and biological targets. *Cell* **2009**, *136*, 731–745. [CrossRef] [PubMed]
24. Barry, K.C.; Ingolia, N.T.; Vance, R.E. Global analysis of gene expression reveals mRNA superinduction is required for the inducible immune response to a bacterial pathogen. *eLife* **2017**, *6*, e22707. [CrossRef] [PubMed]
25. Tiedje, C.; Diaz-Muñoz, M.D.; Trulley, P.; Ahlfors, H.; Laaß, K.; Blackshear, P.J.; Turner, M.; Gaestel, M. The RNA-binding protein TTP is a global post-transcriptional regulator of feedback control in inflammation. *Nucleic Acids Res.* **2016**, *44*, 7418–7440. [CrossRef] [PubMed]
26. Jovanovic, M.; Rooney, M.S.; Mertins, P.; Przybylski, D.; Chevrier, N.; Satija, R.; Rodriguez, E.H.; Fields, A.P.; Schwartz, S.; Raychowdhury, R.; et al. Immunogenetics. Dynamic profiling of the protein life cycle in response to pathogens. *Science* **2015**, *347*, 1259038. [CrossRef] [PubMed]
27. He, K.; Pan, X.; Zhou, H.-R.; Pestka, J.J. Modulation of Inflammatory Gene Expression by the Ribotoxin Deoxynivalenol Involves Coordinate Regulation of the Transcriptome and Translatome. *Toxicol. Sci.* **2013**, *131*, 153–163. [CrossRef] [PubMed]
28. Ceppi, M.; Clavarino, G.; Gatti, E.; Schmidt, E.K.; de Gassart, A.; Blankenship, D.; Ogola, G.; Banchereau, J.; Chaussabel, D.; Pierre, P. Ribosomal protein mRNAs are translationally-regulated during human dendritic cells activation by LPS. *Immun. Res.* **2009**, *5*, 5. [CrossRef] [PubMed]
29. Kitamura, H.; Ito, M.; Yuasa, T.; Kikuguchi, C.; Hijikata, A.; Takayama, M.; Kimura, Y.; Yokoyama, R.; Kaji, T.; Ohara, O. Genome-wide identification and characterization of transcripts translationally regulated by bacterial lipopolysaccharide in macrophage-like J774.1 cells. *Physiol. Genom.* **2008**, *33*, 121–132. [CrossRef] [PubMed]
30. GENCODE. Available online: https://www.gencodegenes.org/ (accessed on 10 October 2017).
31. Quinn, J.J.; Chang, H.Y. Unique features of long non-coding RNA biogenesis and function. *Nat. Rev. Genet.* **2016**, *17*, 47–62. [CrossRef] [PubMed]
32. Mikkelsen, T.S.; Ku, M.; Jaffe, D.B.; Issac, B.; Lieberman, E.; Giannoukos, G.; Alvarez, P.; Brockman, W.; Kim, T.-K.; Koche, R.P.; et al. Genome-wide maps of chromatin state in pluripotent and lineage-committed cells. *Nature* **2007**, *448*, 553–560. [CrossRef] [PubMed]
33. Ulitsky, I.; Bartel, D.P. lincRNAs: Genomics, evolution, and mechanisms. *Cell* **2013**, *154*, 26–46. [CrossRef] [PubMed]
34. Tay, Y.; Rinn, J.; Pandolfi, P.P. The multilayered complexity of ceRNA crosstalk and competition. *Nature* **2014**, *505*, 344–352. [CrossRef] [PubMed]
35. Arthanari, Y.; Heintzen, C.; Griffiths-Jones, S.; Crosthwaite, S.K. Natural antisense transcripts and long non-coding RNA in Neurospora crassa. *PLoS ONE* **2014**, *9*, e91353. [CrossRef] [PubMed]
36. Liu, X.; Li, D.; Zhang, W.; Guo, M.; Zhan, Q. Long non-coding RNA gadd7 interacts with TDP-43 and regulates Cdk6 mRNA decay. *EMBO J.* **2012**, *31*, 4415–4427. [CrossRef] [PubMed]
37. Zhao, J.; Sun, B.K.; Erwin, J.A.; Song, J.-J.; Lee, J.T. Polycomb proteins targeted by a short repeat RNA to the mouse X chromosome. *Science* **2008**, *322*, 750–756. [CrossRef] [PubMed]
38. Milligan, M.J.; Harvey, E.; Yu, A.; Morgan, A.L.; Smith, D.L.; Zhang, E.; Berengut, J.; Sivananthan, J.; Subramaniam, R.; Skoric, A.; et al. Global Intersection of Long Non-Coding RNAs with Processed and Unprocessed Pseudogenes in the Human Genome. *Front. Genet.* **2016**, *7*, 26. [CrossRef] [PubMed]
39. Carrieri, C.; Cimatti, L.; Biagioli, M.; Beugnet, A.; Zucchelli, S.; Fedele, S.; Pesce, E.; Ferrer, I.; Collavin, L.; Santoro, C.; et al. Long non-coding antisense RNA controls Uchl1 translation through an embedded SINEB2 repeat. *Nature* **2012**, *491*, 454–457. [CrossRef] [PubMed]
40. Schein, A.; Zucchelli, S.; Kauppinen, S.; Gustincich, S.; Carninci, P. Identification of antisense long noncoding RNAs that function as SINEUPs in human cells. *Sci. Rep.* **2016**, *6*, 33605. [CrossRef] [PubMed]
41. Kapranov, P.; St. Laurent, G.; Raz, T.; Ozsolak, F.; Reynolds, C.P.; Sorensen, P.H. B.; Reaman, G.; Milos, P.; Arceci, R.J.; Thompson, J.F.; et al. The majority of total nuclear-encoded non-ribosomal RNA in a human cell is "dark matter" un-annotated RNA. *BMC Biol.* **2010**, *8*, 149. [CrossRef] [PubMed]
42. Nadal-Ribelles, M.; Solé, C.; Xu, Z.; Steinmetz, L.M.; de Nadal, E.; Posas, F. Control of Cdc28 CDK1 by a stress-induced lncRNA. *Mol. Cell* **2014**, *53*, 549–561. [CrossRef] [PubMed]
43. Yang, F.; Zhang, H.; Mei, Y.; Wu, M. Reciprocal regulation of HIF-1α and lincRNA-p21 modulates the Warburg effect. *Mol. Cell* **2014**, *53*, 88–100. [CrossRef] [PubMed]

44. St. Laurent, G.; Wahlestedt, C.; Kapranov, P. The Landscape of long noncoding RNA classification. *Trends Genet.* **2015**, *31*, 239–251. [CrossRef]

45. Niazi, F.; Valadkhan, S. Computational analysis of functional long noncoding RNAs reveals lack of peptide-coding capacity and parallels with 3′ UTRs. *RNA* **2012**, *18*, 825–843. [CrossRef] [PubMed]

46. Cabili, M.N.; Trapnell, C.; Goff, L.; Koziol, M.; Tazon-Vega, B.; Regev, A.; Rinn, J.L. Integrative annotation of human large intergenic noncoding RNAs reveals global properties and specific subclasses. *Genes Dev.* **2011**, *25*, 1915–1927. [CrossRef] [PubMed]

47. Wang, Y.; Xue, S.; Liu, X.; Liu, H.; Hu, T.; Qiu, X.; Zhang, J.; Lei, M. Analyses of Long Non-Coding RNA and mRNA profiling using RNA sequencing during the pre-implantation phases in pig endometrium. *Sci. Rep.* **2016**, *6*, 20238. [CrossRef] [PubMed]

48. Derrien, T.; Johnson, R.; Bussotti, G.; Tanzer, A.; Djebali, S.; Tilgner, H.; Guernec, G.; Martin, D.; Merkel, A.; Knowles, D.G.; et al. The GENCODE v7 catalog of human long noncoding RNAs: Analysis of their gene structure, evolution, and expression. *Genome Res.* **2012**, *22*, 1775–1789. [CrossRef] [PubMed]

49. Faghihi, M.A.; Zhang, M.; Huang, J.; Modarresi, F.; Van der Brug, M.P.; Nalls, M.A.; Cookson, M.R.; St. Laurent, G.; Wahlestedt, C. Evidence for natural antisense transcript-mediated inhibition of microRNA function. *Genome Biol.* **2010**, *11*, R56. [CrossRef] [PubMed]

50. Johnsson, P.; Lipovich, L.; Grandér, D.; Morris, K.V. Evolutionary conservation of long non-coding RNAs; sequence, structure, function. *Biochim. Biophys. Acta* **2014**, *1840*, 1063–1071. [CrossRef] [PubMed]

51. Ingolia, N.T.; Lareau, L.F.; Weissman, J.S. Ribosome profiling of mouse embryonic stem cells reveals the complexity and dynamics of mammalian proteomes. *Cell* **2011**, *147*, 789–802. [CrossRef] [PubMed]

52. Van Heesch, S.; van Iterson, M.; Jacobi, J.; Boymans, S.; Essers, P.B.; de Bruijn, E.; Hao, W.; MacInnes, A.W.; Cuppen, E.; Simonis, M. Extensive localization of long noncoding RNAs to the cytosol and mono- and polyribosomal complexes. *Genome Biol.* **2014**, *15*, R6. [CrossRef] [PubMed]

53. Carlevaro-Fita, J.; Rahim, A.; Guigó, R.; Vardy, L.A.; Johnson, R. Cytoplasmic long noncoding RNAs are frequently bound to and degraded at ribosomes in human cells. *RNA* **2016**, *22*, 867–882. [CrossRef] [PubMed]

54. Ruiz-Orera, J.; Messeguer, X.; Subirana, J.A.; Alba, M.M. Long non-coding RNAs as a source of new peptides. *eLife* **2014**, *3*, e03523. [CrossRef] [PubMed]

55. Tran, N.-T.; Su, H.; Khodadadi-Jamayran, A.; Lin, S.; Zhang, L.; Zhou, D.; Pawlik, K.M.; Townes, T.M.; Chen, Y.; Mulloy, J.C.; et al. The AS-RBM15 lncRNA enhances RBM15 protein translation during megakaryocyte differentiation. *EMBO Rep.* **2016**, *17*, 887–900. [CrossRef] [PubMed]

56. Scaria, V.; Hariharan, M.; Maiti, S.; Pillai, B.; Brahmachari, S.K. Host-virus interaction: A new role for microRNAs. *Retrovirology* **2006**, *3*, 68. [CrossRef] [PubMed]

57. Rederstorff, M.; Hüttenhofer, A. Small non-coding RNAs in disease development and host-pathogen interactions. *Curr. Opin. Mol. Ther.* **2010**, *12*, 684–694. [PubMed]

58. Das, K.; Garnica, O.; Dhandayuthapani, S. Modulation of Host miRNAs by Intracellular Bacterial Pathogens. *Front. Cell. Infect. Microbiol.* **2016**, *6*, 79. [CrossRef] [PubMed]

59. Kim, J.K.; Kim, T.S.; Basu, J.; Jo, E.-K. MicroRNA in innate immunity and autophagy during mycobacterial infection. *Cell. Microbiol.* **2017**, *19*. [CrossRef] [PubMed]

60. Zur Bruegge, J.; Einspanier, R.; Sharbati, S. A Long Journey Ahead: Long Non-coding RNAs in Bacterial Infections. *Front. Cell. Infect. Microbiol.* **2017**, *7*, 95. [CrossRef] [PubMed]

61. Ilott, N.E.; Heward, J.A.; Roux, B.; Tsitsiou, E.; Fenwick, P.S.; Lenzi, L.; Goodhead, I.; Hertz-Fowler, C.; Heger, A.; Hall, N.; et al. Long non-coding RNAs and enhancer RNAs regulate the lipopolysaccharide-induced inflammatory response in human monocytes. *Nat. Commun.* **2014**, *5*, 3979. [CrossRef] [PubMed]

62. Leong, H.S.; Dawson, K.; Wirth, C.; Li, Y.; Connolly, Y.; Smith, D.L.; Wilkinson, C.R.M.; Miller, C.J. A global non-coding RNA system modulates fission yeast protein levels in response to stress. *Nat. Commun.* **2014**, *5*, 3947. [CrossRef] [PubMed]

63. Koirala, P.; Huang, J.; Ho, T.-T.; Wu, F.; Ding, X.; Mo, Y.-Y. LncRNA AK023948 is a positive regulator of AKT. *Nat. Commun.* **2017**, *8*, 14422. [CrossRef] [PubMed]

64. Li, P.; Xue, W.-J.; Feng, Y.; Mao, Q.-S. Long non-coding RNA CASC2 suppresses the proliferation of gastric cancer cells by regulating the MAPK signaling pathway. *Am. J. Transl. Res.* **2016**, *8*, 3522–3529. [PubMed]

65. Li, R.; Zhang, L.; Jia, L.; Duan, Y.; Li, Y.; Bao, L.; Sha, N. Long non-coding RNA BANCR promotes proliferation in malignant melanoma by regulating MAPK pathway activation. *PLoS ONE* **2014**, *9*, e100893. [CrossRef] [PubMed]

66. Bhattacharyya, S.; Vrati, S. The Malat1 long non-coding RNA is upregulated by signalling through the PERK axis of unfolded protein response during flavivirus infection. *Sci. Rep.* **2015**, *5*, 17794. [CrossRef] [PubMed]

toxins

MDPI

Article

Evidence for Complex Formation of the *Bacillus cereus* Haemolysin BL Components in Solution

Franziska Tausch [1], Richard Dietrich [1], Kristina Schauer [1], Robert Janowski [2], Dierk Niessing [2,3], Erwin Märtlbauer [1] and Nadja Jessberger [1,*]

[1] Department of Veterinary Sciences, Faculty of Veterinary Medicine, Ludwig-Maximilians-Universität München, Schönleutnerstr 8, 85764 Oberschleißheim, Germany; Franziska.Tausch@gmx.de (F.T.); r.dietrich@mh.vetmed.uni-muenchen.de (R.D.); kristina.schauer@mh.vetmed.uni-muenchen.de (K.S.); e.maertlbauer@mh.vetmed.uni-muenchen.de (E.M.)

[2] Institute of Structural Biology, Helmholtz Zentrum München-German Research Center for Environmental Health, Ingolstädter Landstr. 1, 85764 Neuherberg, Germany; robert.janowski@helmholtz-muenchen.de (R.J.); niessing@helmholtz-muenchen.de (D.N.)

[3] Biomedical Center of the Ludwig-Maximilians-Universität München, Department of Cell Biology, 82152 Planegg-Martinsried, Germany

* Correspondence: n.jessberger@mh.vetmed.uni-muenchen.de; Tel.: +49-89-2180-78577; Fax: +49-89-2180-78602

Academic Editor: Alexey S. Ladokhin
Received: 16 August 2017; Accepted: 12 September 2017; Published: 16 September 2017

Abstract: Haemolysin BL is an important virulence factor regarding the diarrheal type of food poisoning caused by *Bacillus cereus*. However, the pathogenic importance of this three-component enterotoxin is difficult to access, as nearly all natural *B. cereus* culture supernatants additionally contain the highly cytotoxic Nhe, the second three-component toxin involved in the aetiology of *B. cereus*-induced food-borne diseases. To better address the toxic properties of the Hbl complex, a system for overexpression and purification of functional, cytotoxic, recombinant (r)Hbl components L_2, L_1 and B from *E. coli* was established and an *nheABC* deletion mutant was constructed from *B. cereus* reference strain F837/76. Furthermore, 35 hybridoma cell lines producing monoclonal antibodies (mAbs) against Hbl L_2, L_1 and B were generated. While mAbs 1H9 and 1D8 neutralized Hbl toxicity and thus, represent important tools for future investigations of the mode-of-action of Hbl on the target cell surface, mAb 1D7, in contrast, even enhanced Hbl toxicity by supporting the binding of Hbl B to the cell surface. By using the specific mAbs in Dot blots, indirect and hybrid sandwich enzyme immuno assays (EIAs), complex formation between Hbl L_1 and B, as well as L_1 and L_2 in solution could be shown for the first time. Surface plasmon resonance experiments with the rHbl components confirmed these results with K_D values of 4.7×10^{-7} M and 1.5×10^{-7} M, respectively. These findings together with the newly created tools lay the foundation for the detailed elucidation of the molecular mode-of-action of the highly complex three-component Hbl toxin.

Keywords: *Bacillus cereus*; complex formation; enterotoxins; haemolysin BL; monoclonal antibodies

---

1. Introduction

The Gram-positive, facultative anaerobe and rod-shaped bacterium *Bacillus cereus* has become increasingly important as a cause of food poisoning outbreaks. It is ubiquitous in soil, sediments, dust and plants [1]. From there, it can easily be spread to different foods. *B. cereus* is found in an extraordinary variety of food, as for example in milk and dairy products, rice and pasta or spices, dry foods and vegetables [1–3]. Food poisoning caused by *B. cereus* is mostly moderate and self-limiting, but also severe and even lethal cases have been reported [4–6].

B. cereus causes mainly two types of foodborne diseases, an emetic and a diarrheal form. The first is characterized by nausea and vomiting and is caused by the cyclic dodecadepsipeptide cereulid, which is produced in foods before consumption [7,8]. Three types of enterotoxins are responsible for the diarrheal form. These are produced foremost in the intestine from viable bacteria that, most likely as spores, survived the stomach passage [9]. These enterotoxins are the three component complexes Hbl (haemolysin BL; [10]) and Nhe (non-haemolytic enterotoxin; [11]), and the single protein CytK (cytotoxin K; [4]). CytK has been reported to be haemolytic, cytotoxic, necrotic and is a member of the family of β-barrel pore forming toxins [12,13]. Only very few strains express the highly toxic CytK1 variant and are classified as *Bacillus cytotoxicus* [14].

The more complex Nhe and Hbl toxins consist of three components each, namely NheA, B and C and Hbl L_2, L_1 and B. It has been shown that for both toxins all three components are needed for maximum biological activity [10,15]. Intensive studies have been performed on the mode of action of Nhe. It has been shown that NheB and C, but not NheA can bind to Vero cells. Nhe is a pore forming toxin inducing cell lysis with an optimum molar ratio for maximum toxicity of A:B:C = 10:10:1. Increasing ratios of NheC lead to inhibition of the toxic activity [15,16]. Moreover, a specific binding order, i.e., NheC-B-A, is necessary for Nhe activity [17–19]. The mode-of-action of Nhe was clarified with the help of monoclonal antibodies (mAbs), which hinder the interaction between the single Nhe components and thus, neutralize the toxic activity [18,20]. On the other hand, these specific mAbs were also used to detect complex formation of Nhe components in culture supernatants, such as the interaction between NheB and C [21].

Hbl was originally purified from *B. cereus* strain F837/76 and a binding component (B) as well as two lytic components (L_2 and L_1) were identified [22,23]. Several different ideas about the mode of action of Hbl exist. First, it has been suggested that each component is able to bind individually to erythrocytes and thus, that they assemble into a "membrane attack complex", then form a transmembrane pore and lyse the cells [24]. Osmotic protection assays showed that Hbl is a pore forming toxin and that pores are smaller than 1.2 nm [24]. On blood agar plates, it was observed that excess of Hbl L_1 or B inhibits haemolytic activity [24,25]. Thus, a certain concentration ratio of the Hbl components might also be required for maximum activity, but this is, in contrast to Nhe, so far unknown. In another approach it was shown that sequential binding in the specific binding order of Hbl B-L_1-L_2 leads to toxic activity in cell viability tests on Chinese hamster ovary cells [19]. As the crystal structure of Hbl B was solved, a high structural similarity to *E. coli* haemolysin E (HlyE; ClyA) was observed despite low sequence homology [26]. Based on this similarity, another model was proposed in which Hbl B alone might be able to oligomerize on the cell surface and form a pore. L_2 and L_1, which are definitely required for toxic activity, might either stabilize B, induce conformational changes to B or even enter the cell [26].

Investigating Hbl activity in natural *B. cereus* culture supernatants is difficult, as all known enteropathogenic strains that bear *hbl* also have the *nhe* operon. The *nhe* genes can be found in all enteropathogenic *B. cereus* strains, the *hbl* genes in about 45%–65% [27–29]. Early studies on Hbl activity were carried out with proteins purified via anion exchange chromatography from *B. cereus* culture supernatants [23,30]. These might have contained trace contaminations and the use of recombinantly expressed and purified Hbl (rHbl) components was recommended [10]. For a long time rHbl components could not be generated, it was even suggested that the individual components might be toxic for *E. coli* [31]. Only in 2013, Sastalla and co-workers were able to overexpress *B. cereus* Nhe and Hbl proteins in a *Bacillus anthracis* expression system [19].

The aim of this study was to create suitable and effective tools by which the complex mode-of-action of Hbl can be investigated. First, functional rHbl components were overexpressed and purified from *E. coli* in an easy, fast and secure (S1) system. Secondly, a deletion mutant from *B. cereus* reference strain F837/76 was constructed lacking the *nhe* operon. Finally, a whole set of highly specific mAbs against the three Hbl components was generated. Based on these reagents, it could be shown for the first time that Hbl components L_1 and B, as well as L_1 and L_2 form complexes in solution.

2. Results

2.1. Generation of Functional Recombinant Hbl (rHbl) Components and a ΔnheABC Mutant

To generate functional, recombinant Hbl toxin, an approach based on the Strep-tag® purification system (iba lifesciences) was used. rHbl L_2 and B were expressed with an N-terminal, rHbl L_1 with a C-terminal strep-tag. Sequences encoding putative secretion signal peptides were eliminated. Amino acid sequences and predicted molecular weight of the new recombinant proteins are shown in Figure 1A. The proteins were overexpressed in *E. coli* BL21 (DE3) and purified via affinity chromatography. Purification was controlled in Western blots (Figure 1B). Each of the single components was detected by a strep-specific antibody as well as by the Hbl-specific mAbs. rHbl L_2 and L_1 were detected at their predicted molecular weight of 48.7 and 42.4 kDa, while rHbl B appeared slightly bigger in Western blot analyses than its predicted 40.9 kDa. For rHbl L_1, a second band was detected with the L_1- but not with the strep-specific mAb, indicating partial cleavage of the tag.

A ΔnheABC mutant was constructed from *B. cereus* strain F837/76. The *nhe* operon was replaced by a spectinomycin resistance cassette, which was confirmed by sequencing. The start and stop codon, as well as the first 36 and the last 18 bp of the operon remained. No phenotypical differences could be detected compared to the wild type (Figure 1C). Western blots (Figure 1D) confirmed that no Nhe toxin was produced by the deletion strain.

To determine the toxic activity of the newly generated rHbl components and the ΔnheABC strain, they were analysed by PI influx tests as well as by WST-1-bioassays. The rHbl components were clearly able to induce pore formation in Vero cells when applied simultaneously (Figure 1E). Fluorescence could be measured after approximately 25 min, while supernatant of strain F837/76, which was used as control, caused almost immediate PI influx. Simultaneously applied rHbl components were also able to kill Vero cells (Figure 1F), although the reciprocal titre (the dilution to get 50% viable cells after 24 h) was significantly lower than that of F837/76 supernatant (250 compared to 700). The ΔnheABC mutant showed decelerated PI influx compared to the wild type (Figure 1E) as well as a reduced reciprocal cytotoxicity titre (400 compared to 700; Figure 1F). Altogether, these results prove that, in contrary to earlier claims, overexpression and purification of functional rHbl in *E. coli* as well as the deletion of the *nhe* operon in *B. cereus* is indeed possible.

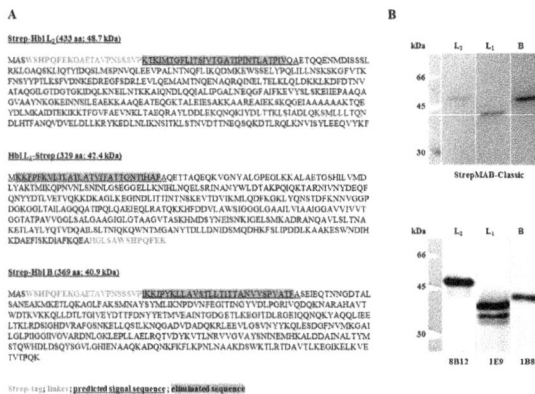

Figure 1. *Cont.*

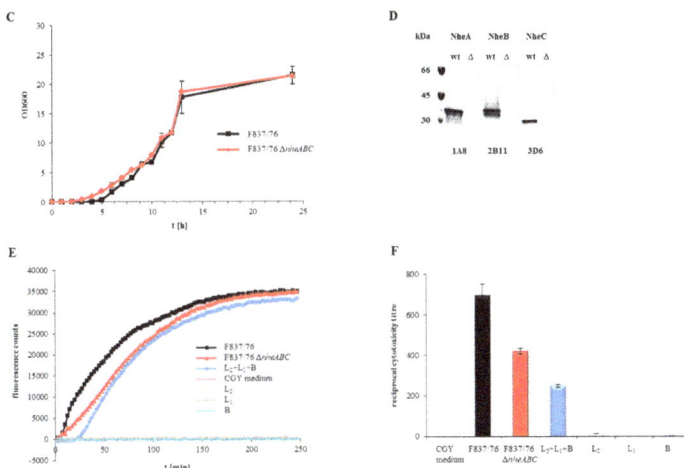

Figure 1. Properties of the recombinant Hbl components and the *nheABC* deletion strain. (**A**) Amino acid sequences of rHbl L_2, L_1 and B. Strep-tags are shown in blue, linker amino acids in orange letters, the predicted secretion signal sequences are underlined and eliminated sequences are highlighted in grey. From these sequences the molecular weight was predicted (http://web.expasy.org/compute_pi/). (**B**) Detection of the rHbl components in Western blots. The strep-specific StrepMAB-Classic was used for detection (upper blot), as well as the Hbl-specific mAbs 8B12 (L_2) [32], 1E9 (L_1) [29] and 1B8 [29] (lower blot). (**C**) Growth of *B. cereus* strains F837/76 and F837/76 $\Delta nheABC$ in LB medium. Medium was inoculated to an OD_{600} of 0.005 and growth at 37 °C was monitored for 24 h. (**D**) Western blot. NheA, B and C were detected in the supernatant of strain F837/76 using the mAbs 1A8 [20], 2B11 [20] and 3D6 [21], respectively. These proteins were not detected in the supernatant of the *nheABC* deletion mutant. (**E**) Influx of PI into Vero cells, represented by increasing fluorescence counts. 1.5 pmol/µL rHbl components were either used separately or mixed in a 1:1:1 ratio and applied in 1:40 dilution to the cells, as was supernatant of F837/76 $\Delta nheABC$. CGY medium and supernatant of F837/76 were used as controls. (**F**) Results of a WST-1-bioassay (Vero cells) determining toxic activity of the *nheABC* deletion strain and the rHbl components (1.5 pmol/µL each, separately or mixed in a 1:1:1 ratio). CGY medium and supernatant of F837/76 were used as controls.

2.2. Generation of Monoclonal Antibodies (mAbs) against Hbl

Highly specific mAbs were generated in this study according to established procedures [20,32]. For immunogen preparation, strain MHI 1532 was chosen, as it showed the highest Hbl B titres in indirect EIAs and the highest Hbl L_2 titres in sandwich EIAs of all strains tested in preliminary experiments (data not shown). The immunogen was gained by purification of the culture supernatant via Hbl B-specific IAC with the already established Hbl B-specific mAb 1B8 [29]. Purified Hbl toxin was used for immunization and two booster injections of mice. After cell fusion, 35 hybridoma cell lines secreting Hbl-reactive antibodies could be identified (Table S1). The target antigens of the mAbs were determined in indirect EIAs using rHbl components. Twenty-nine hybridoma cell lines produced Hbl B-specific mAbs, four mAbs were cross-reactive with Hbl B and Hbl L_1, and two mAbs were specific for Hbl L_2. As mice were immunized with a preparation of IAC-purified Hbl B from *B. cereus* culture supernatants (see above), this finding gave a first hint that the single Hbl components form complexes in solution.

With respect to affinity, stability and productivity of the hybridoma cell lines, 1G8 and 12D12 (Hbl B/L_1), 1D8 and 1H9 (Hbl L_2), and 2G4, 1D7, 1D12 and 1C12 (Hbl B) were chosen for mass production and the respective, purified mAbs were used for further experiments.

2.3. The Generated mAbs Show Neutralizing and Enhancing Properties towards Hbl Toxicity

To investigate the neutralizing properties of the generated mAbs against Hbl, antibodies were applied simultaneously with culture supernatants of *B. cereus* strain F837/76 Δ*nheABC* in WST-1-bioassays on Vero cells. The reciprocal titre of the untreated supernatant was set to 100% and relative cytotoxicity of all tested samples was compared to that value. mAbs 1H9 and 1D8 (Hbl L$_2$-specific) clearly reduced the toxic activity of the Hbl-containing *B. cereus* supernatant by approximately 60%, while unexpectedly mAbs 1D7 (Hbl B), 1D12 (Hbl B) and 12D12 (Hbl B/L$_1$) enhanced it by 256, 101 and 165%, respectively (Figure 2A). Isotype controls as well as all other tested mAbs showed no significant influence on Hbl toxicity. The three mAbs enhancing Hbl toxicity were used in flow cytometric analyses to determine their influence on Hbl B binding to Vero cells. For that, Vero cells were incubated for 1 h with rHbl B (6.5 pmol/mL) and then probed with Alexa Fluor® 488-labelled mAb 1G8 (Hbl B/L$_1$-specific). These settings resulted in approximately 42% fluorescence (FL1)-positive cells (Table 1). Co-incubation of rHbl B and mAbs 1D12 and 12D12 on Vero cells resulted in no and only slightly enhanced number of FL1-positive cells, respectively. However, when rHbl B was co-incubated on Vero cells with the Hbl B-reactive mAb 1D7, nearly all cells were FL1 positive (96.37%) (Table 1 and Figure 2B). Under consecutive incubation conditions, in which Vero cells were first incubated with rHbl B and then, after washing, with mAb 1D7, this effect was not seen. Negative and isotype controls proved the specificity of the reaction (Table 1). Overall, it could be demonstrated that rHbl B binds specifically to Vero cells and that Hbl B-specific mAb 1D7 enhances this binding by a so far unknown mechanism. The variety of enhancing or neutralizing properties indirectly indicates that the generated mAbs recognize different epitopes.

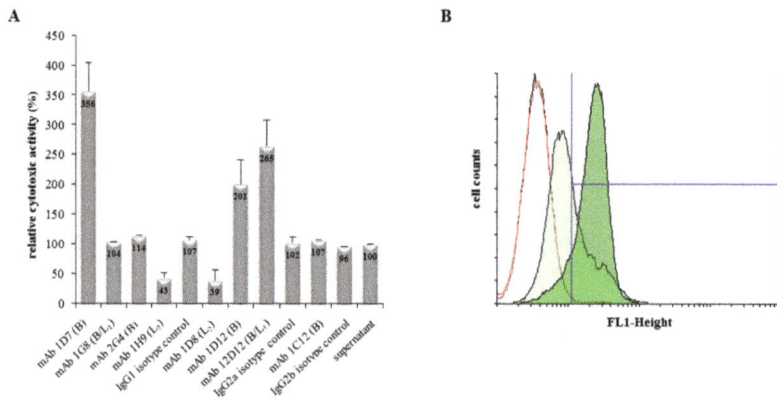

Figure 2. Influence of mAbs on Hbl toxicity. (**A**) WST-1-Bioassay on Vero cells. Supernatant of F837/76 Δ*nheABC* was applied as serial dilution. mAb (10 µg/well) was added to each dilution and incubated with the cells for 24 h. Cytotoxicity titres were determined by addition of WST-1. The reciprocal titre of the untreated *B. cereus* supernatant was set to 100%. (**B**) Flow cytometry results of Vero cells treated with rHbl B. Cells were incubated for 1 h with either only buffer (black curve), Hbl B-specific mAb 1D7 (red curve), rHbl B (light green filled) or rHbl B + mAb 1D7 (dark green filled). Cell-bound rHbl B was detected by using Alexa Fluor® 488-labelled mAb 1G8 (Hbl B-specific).

Table 1. Flow cytometry of Vero cells treated with rHbl B. Vero cells (1×10^6) were incubated for 1 h with rHbl B (6.5 pmol/mL) and simultaneously with mAbs 1D7 (Hbl B-specific), 1D12 (Hbl B-specific) or 12D12 (Hbl B/L_1-specific) (ratio mAb:rHbl B = 1:1). Fluorescence was detected via Hbl B-specific Alexa Fluor® 488-labelled mAb 1G8 (FL1, fluorescence at 488 nm).

Sample	FL1-Positive (%)
negative controls:	
-	0.69 ± 0.29
1D7	0.96 ± 0.12
1D12	0.68 ± 0.03
12D12	0.53 ± 0.06
rHbl B and mAbs co-incubated:	
rHbl B	41.95 ± 6.3
rHbl B + 1D7	96.37 ± 4.34
rHbl B + 1D12	43.99 ± 6.42
rHbl B + 12D12	50.81 ± 4.11
rHbl B and mAb 1D7 consecutively:	
rHbl B, 1D7	32.76 ± 24.61
isotype controls:	
rHbl B + IgG1	58.37 ± 7.33
rHbl B + IgG2a	46.36 ± 1.34

2.4. Complex Formation of rHbl Components

The panel of Hbl-specific mAbs with putative different epitope-specificity offered the opportunity to investigate possible interactions between the single rHbl components. For that, Dot blots and enzyme immunoassays were performed. For Dot blot analyses, a dilution series (480–3.75 pmol) of the first rHbl component was applied to a PVDF membrane, which was then incubated with the second rHbl component (30 pmol in PBS). For detection, a specific mAb against the second rHbl component was used. Hereby, specific interactions between rHbl L_1 and B as well as rHbl L_1 and L_2, but not rHbl L_2 and B were demonstrated, whereat the application order of the rHbl components was not important (Figure 3A and Figure S1). A similar composition applied in indirect EIAs confirmed the specific complex formation between rHbl L_1 and B, and rHbl L_1 and L_2 (Figure 3B). These complexes could also be detected in a different approach where rHbl L_1 and B, and rHbl L_1 and L_2 were pre-incubated for 30 min and subsequently detected in highly specific sandwich EIAs (Figure 4A). This variant also allowed the comparison of different ratios of the single rHbl components (Figure 4B). Hbl complexes were detected in all tested variants. In the 1E9-1B8-sandwich EIA, rHbl L_1 + B complexes were less well detected with excess of B. Excess of L_1 had little influence on complex detection compared to the 1:1 ratio. In the 1G8 (Hbl HblB/L_1-1H9 Hbl L_2)-sandwich EIA, rHbl $L_1 + L_2$ complexes were less well detected with excess of L_2. A ratio of L_1:L_2 of 5:1 resulted in the highest absorption values, while further excess (10:1) showed no difference to the 1:1 control. Complex formation between rHbl L_2 and B could be detected neither in indirect nor in sandwich EIAs (data not shown).

Finally, complex formation between rHbl L_1 and L_2 as well as rHbl L_1 and B was confirmed via surface plasmon resonance (SPR). For that, rHbl L_1 was immobilized, and rising concentrations (from 7.8 nM to 2 µM) of rHbl L_2 and B were added. K_D values of 1.5×10^{-7} M and 4.7×10^{-7} M were determined, respectively (Figure 5A,B). When rHbl L_2 was coupled to the sensor chip, only interaction with rHbl L_1 was detected, not with B (data not shown). Interestingly, switching rHbl B and L_2 (B coupled and L_2 used as ligand), resulted in specific interaction of the two proteins, with a, compared to the results obtained in Figure 5A,B, relative low K_D value of 3.4×10^{-6} M (Figure 5C).

A

B

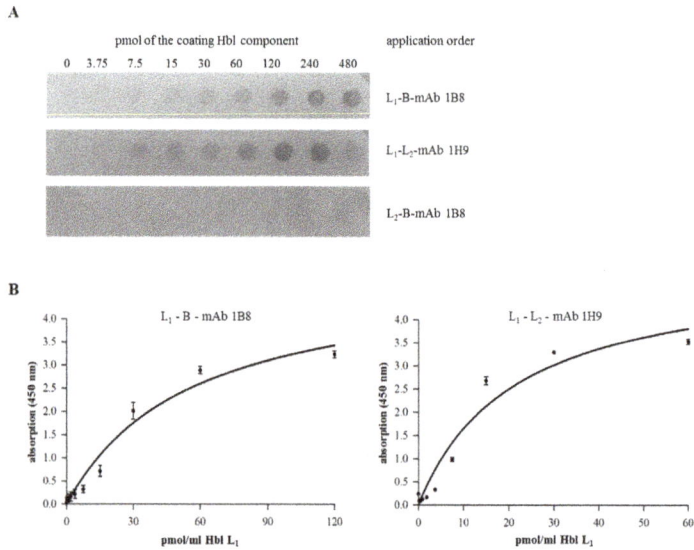

Figure 3. Detection of rHbl complex formation. (**A**) Dot blot. PVDF membranes were coated with rising concentrations (3.75–480 pmol) of different rHbl components. After blocking, the membrane was incubated in PBS with the second component (30 pmol). Proteins were detected using the Hbl B-specific mAb 1B8 [29] and the Hbl L_2-specific mAb 1H9 (this study). Inversion of the protein order showed similar results and negative controls confirmed the specificity of the reaction (see Figure S1). (**B**) Indirect EIA. The first rHbl component was applied as serial dilution to a microtiter plate. After washing, the second rHbl component was applied in constant concentration (60 pmol/mL). After blocking, Hbl B-specific mAb 1B8 [29] and Hbl L_2-specific mAb 1H9 (this study) were applied, respectively, followed by rabbit-anti-mouse-HRP conjugate for detection. Details on the non-linear regression are shown in Table S2.

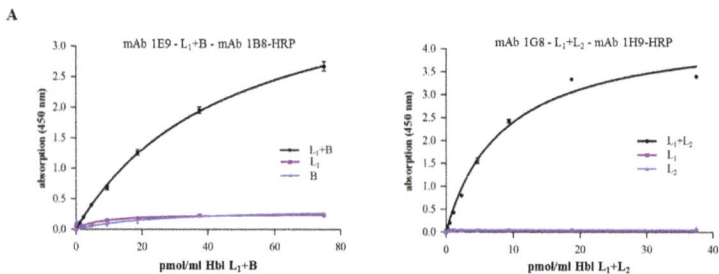

A

Figure 4. *Cont.*

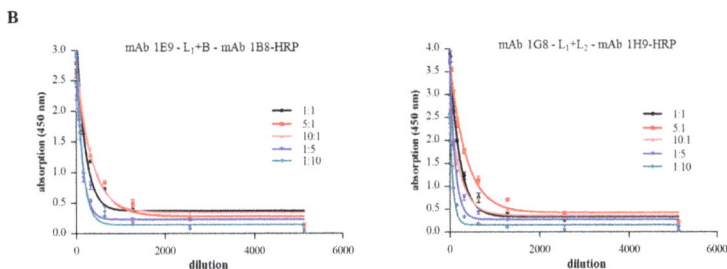

Figure 4. Detection of rHbl complex formation via sandwich EIAs. (**A**) Microtiter plates were coated with mAbs 1E9 (Hbl L_1) [29] and 1G8 (Hbl B/L_1) (this study), respectively. rHbl components were pre-mixed (each 1.5 pmol/μL in an 1:1 ratio), incubated for 30 min at RT and, after blocking, applied as serial dilution. After washing, the specific conjugates 1B8-HRP (Hbl B) [29] and 1H9-HRP (Hbl L_2) (this study) were used for detection. (**B**) In an analogous approach, rHbl components were pre-mixed in different concentration ratios (each 1.5 pmol/μL, ratios 1:1, 5:1, 10:1, 1:5 and 1:10). Details on the non-linear regression are shown in Table S2.

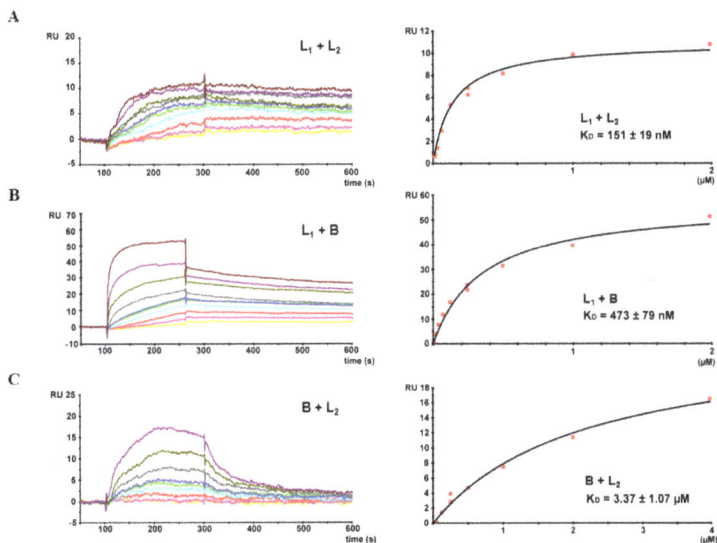

Figure 5. SPR measurement of the rHbl components. On the left side, a representative sensogram is depicted. Right panels show binding curves fitted to a one-site-binding model with calculated equilibrium dissociation rates (K_D) and errors as standard deviations from three or four independent experiments. (**A**) rHbl L_1 immobilized on a sensor chip. Concentration series (7.8 nM—yellow, 15.6 nM—magenta, 31.2 nM—red, 62.5 nM—cyan, 125 nM—green, 250 nM—blue and gray, 500 nM—dark yellow, 1 μM—dark magenta and 2 μM—dark red) of rHbl L_2 were applied. (**B**) rHbl L_1 immobilized on a sensor chip. Concentration series (7.8 nM—yellow, 15.6 nM—magenta, 31.2 nM—red, 62.5 nM—cyan, 125 nM—green, 250 nM—blue and gray, 500 nM—dark yellow, 1 μM—dark magenta and 2 μM—dark red) of rHbl B were applied. (**C**) rHbl B immobilized on a sensor chip. Concentration series (31.2 nM—yellow, 62.5 nM—magenta, 125 nM—red, 250 nM—cyan and green, 500 nM—blue, 1 μM—gray, 2 μM—dark yellow and 4 μM—dark magenta) of rHbl L2 were applied. RU: response units. Details on the non-linear regression are shown in Table S2.

2.5. Complex Formation in Natural B. cereus Culture Supernatants

To detect complex formation of Hbl components in natural *B. cereus* supernatants, sandwich EIAs were performed with serial dilutions of F837/76 supernatant (analogously to Figure 4). Hbl L_1-B complexes could be detected (Figure 6A). As only very weak signals were detected in the Hbl L_1-L_2-specific sandwich EIA (Figure 6B), the specific mAbs were switched (anti-L_2 1H9 as capture and anti-L_1 1G8 as detection antibody) and clearer signals were determined (Figure 6C). Altogether, with the complex sandwich EIAs, effective tools were developed to prove Hbl complex formation not only within recombinant Hbl components, but also in natural *B. cereus* culture supernatants.

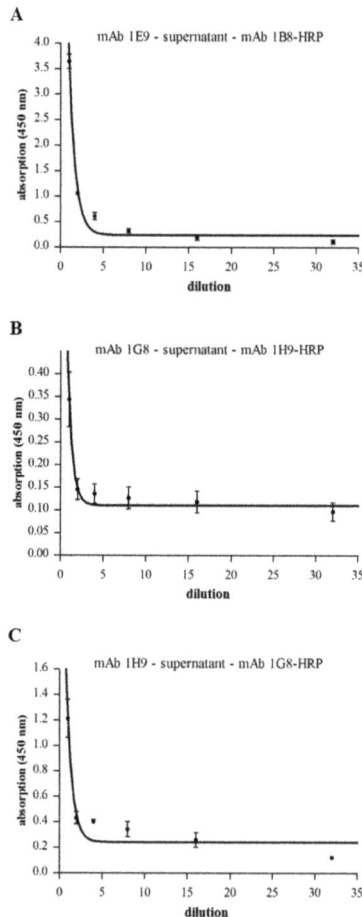

Figure 6. Hbl complex formation in natural *B. cereus* supernatants. In all tests, supernatant of strain F837/76 was applied as serial dilution. (**A**) Sandwich EIA specific for Hbl L_1-B complexes, using mAb 1E9 (Hbl L_1) and the 1B8-HRP conjugate (Hbl B) [29]. (**B**) Sandwich EIA specific for Hbl L_1-L_2 complexes, using mAb 1G8 (Hbl L_1) and the 1H9-HRP conjugate (Hbl L_2) (this study). (**C**) Due to weak results in Figure 6B, the capture and detection mAbs for Hbl L_1 (1G8) and L_2 (1H9) were switched. Details on the non-linear regression are shown in Table S2.

In an independent approach, MHI 1532 supernatant was purified via IAC using the Hbl B-specific mAb 1B8 [29] equivalently to the generation of the immunogen. Single fractions were collected and tested in sandwich EIAs. Results of three independent approaches are summarized in Table 2. Hbl B and L_1 bound with high capacity to the IAC column, less than 5% of these toxin components were detected in sample flow-through and wash fraction. The detectable proportion of Hbl B and L_1 in the elution fraction averaged at 187% and 180% compared to the MHI 1532 supernatant. The recovery rate of Hbl L_2 in the elution fraction averaged at only 29%, while 51% were detected in the sample flow-through and 20% in the wash fraction. A sandwich EIA against the toxin component NheB was performed to exclude unspecific binding of toxin components to the immunosorbent. Indeed, 97% of NheB were detected in the sample flow-through. As a control, the identical experiment was performed using the Hbl L_2-specific mAb 1H9 coupled to the CnBr-activated sepharose 4B matrix. Ninety percent of Hbl L_2 were retrieved in the elution fraction, whereas only 2% and 1% of Hbl L_1 and B were detected, respectively (Table 2). Corresponding Western blot analyses confirmed the foregoing results (Figure 7). Altogether, Hbl B as well as L_2 and L_1 could be detected in the elution fraction of the Hbl B-specific 1B8-IAC and prove the pronounced complex formation between the single Hbl components in natural supernatants.

Table 2. Proportion of Hbl components L_2, L_1 and B in sample flow-through, wash and elution fraction after IAC purification of *B. cereus* MHI 1532 supernatant. Hbl B-specific mAb 1B8 [29] and Hbl L_2-specific mAb 1H9 (this study) were used for IAC. Hbl components were determined in sandwich EIAs: Hbl L_2: mAbs 1A12 [32]/1H9-HRP (this study); Hbl L_1: mAbs 1E9 [29]/1G8-HRP (this study); Hbl B: mAbs 1D12 (this study)/1B8-HRP [29]. NheB was also measured as a control (mAbs 2B11/1E11-HRP) [20]. EIA titres were multiplied with the respective sample volume resulting in units. The respective units determined in MHI 1532 supernatant were set to 100%.

Hbl B-Specific IAC (1B8)			
	Sample Flow-Through	Wash Fraction	Elution Fraction
Hbl L_2	51%	20%	29%
Hbl L_1	<5%	<5%	180%
Hbl B	<5%	<5%	187%
NheB	97%	-	3%

Hbl L_2-Specific IAC (1H9)			
	Sample Flow-Through	Wash Fraction	Elution Fraction
Hbl L_2	9%	1%	90%
Hbl L_1	55%	4.5%	2%
Hbl B	110%	10%	1%

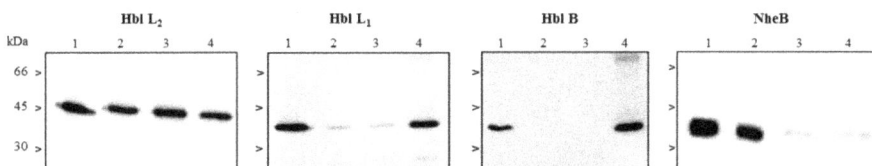

Figure 7. Western blot analyses of Hbl B-specific IAC fractions of MHI 1532 supernatant: 1, supernatant of strain MHI 1532; 2, sample flow-through; 3, wash fraction; and 4, elution fraction. Toxin components were detected using the following mAbs: 1H9 (Hbl L2), 1E9 (Hbl L1), 1B8 (Hbl B), and 1E11 (NheB) ([20,33] and this study).

3. Discussion

With the construction of an *nheABC* deletion mutant and the establishment of a system for overexpression and purification of active recombinant Hbl components in *E. coli*, the first important steps towards clarification of the molecular mode-of-action of Hbl have been made. Earlier studies had been laborious and not fully conclusive, as the second three-component enterotoxin Nhe is always present in crude *B. cereus* culture supernatants [28,29]. Furthermore, Hbl was purified in these early studies via anion exchange chromatography, including the possibility that the resulting preparations might have contained trace contaminations of Nhe [10,23,30]. As Nhe dominates toxicity studies, another approach to assess the peculiar Hbl activity was to remove NheB via subtractive immunoaffinity chromatography (IAC) and attribute the remaining toxic activity to Hbl [33]. By using rHbl components in the present study, no other *B. cereus* toxin is involved and the single Hbl components can be applied in defined concentrations and proportions. The three rHbl components produced in *E. coli* are functionally active and cytotoxic, but it must be admitted that cytotoxic activity is reduced compared to natural Hbl appearing in *B. cereus* culture supernatants (refer to Figure 1C,D).

Based on these reactive proteins and a broad range of Hbl-specific mAbs, complex formation between the single Hbl components could be detected for the first time. Via Dot blots, hybrid sandwich EIAs and immunoaffinity chromatography, Hbl L_1-B and Hbl L_1-L_2 complex formation could be shown both for the recombinant Hbl proteins as well as in natural *B. cereus* culture supernatants (refer to Figures 3, 4 and 6). Particularly, the L_1 and B components seem to be highly complexed. The only, comparably weak interaction between Hbl B and L_2 was observed in SPR measurements, when rHbl B was immobilized and high concentrations of rHbl L_2 were added (refer to Figure 5C). Apart from that, SPR measurements confirmed the results obtained in hybrid sandwich EIAs, indicating that there is an interaction and complex formation between Hbl L_1 and L_2 as well as L_1 and B. This Hbl complex formation also explains why all three Hbl components were detected in the elution fraction of Hbl B-specific IAC from strain MHI 1532 (Table 2 and Figure 7), which was initially used for immunogen preparation, and why mAbs specific for all three Hbl components (Table S1) were generated. Thus, it can be postulated that in natural *B. cereus* culture supernatants Hbl B is to a large extent bound in complexes. Further on, these complexes are characterized by K_D values of 4.7×10^{-7} M (L_1-B) and 3.4×10^{-6} M (B-L_2) (Figure 5). These are relatively low constants compared to the situation found for the NheB-C complexes, for which a value of 4.8×10^{-10} M was found [34]. This might be the reason why single Hbl components were obtained from *B. cereus* culture supernatants with the applied classical chromatographic approach in the early studies [23,30].

According to sequence homologies between the two three-component enterotoxin complexes of *B. cereus*, Hbl L_2 correlates with NheA, Hbl L_1 with NheB and Hbl B with NheC [31,35]. Thus, the Hbl L_1-B complexes found in this study would correspond to NheB–NheC complexes, which have already been well-described. It was shown that a defined level of NheB–NheC complexes as well as a sufficient amount of free NheB is necessary for efficient cell binding and toxicity [21,34]. Based on these data, a detailed model on the mode-of-action of Nhe could be developed [36]. Assuming that Hbl L_1 is equivalent to NheB and Hbl B to NheC, it can be speculated that the Hbl L_1-B complexes alone might be able to bind to the target cells and form a kind of "pro-pore" [34]. Analogously to NheA, Hbl L_2 would be the third component which, presumably after undergoing conformational changes, binds to the complex and completes the pore. However, the comparably low K_D values of the complexes found in this study, as well as the proposed binding order of the single components B-L_1-L_2 [19] oppose this theory.

Interestingly, in our study some of the newly generated mAbs against Hbl B enhanced Hbl toxicity towards Vero cells (refer to Figure 2A). By using purified rHbl B and flow cytometry analyses, an increased binding of Hbl B to Vero cells in the presence of the Hbl B-specific mAb 1D7 was found (Figure 2B and Table 1). Various studies describe enhancing effects of virus-specific mAbs, known as antibody-dependent-enhancement (ADE) [37,38]. ADE has been described for Dengue fever [39], HIV [40], Feline Infectious Peritonitis [41] and Ebola [42]. In 2004, ADE was first observed regarding the lethal toxin of *B. anthracis* [43]. The authors postulated stabilization of PA (protective antigen) binding to the cell surface via interaction of mAbs with macrophage Fcγ receptors. In contrast to this, involvement of Fc receptors can be excluded in our study, as they do not occur on the surface of Vero cells. Thus, the reason for the increased rHbl B binding remains unclear. One can speculate that through the two binding sites of the mAb, an antibody-mediated cross-linking of rHbl B is induced, which might promote the binding of rHbl B to the cell surface. It is further possible that binding of the mAb results in a change of the tertiary structure of rHbl B enabling a better binding to the target cells.

Besides these toxicity-enhancing mAbs, also mAbs with neutralizing properties were also found. This is particularly true for the mAbs reactive with Hbl L_2 (refer to Table S1 and Figure 2A). However, these mAbs neutralize Hbl toxicity only by approximately 60%, in contrast to an earlier study, in which an NheB-specific mAb (1E11) was generated that neutralizes Nhe-dependent toxicity almost completely [20]. The partial neutralization of Hbl L_2-specific 1H9 and 1D8 indicates that both mAbs interact with regions of the toxin which are important, but not mandatory for toxic activity. Further studies are necessary to determine whether these mAbs hinder the proposed conformational change of the Hbl components during pore formation [26] or the putative attachment of Hbl L_2 to cell-bound Hbl B/L_1. It might also be possible that the mAbs are not able to access Hbl L_2, as it is partially bound in complexes with the other components, as shown in this study.

The demonstration of Hbl complex formation in the current study is a first step to characterize its mode-of-action in detail. Further studies are in progress to elucidate the epitope specificity of the mAbs to get new insights into the interaction of the single components. Cell culture studies will follow to investigate the impact of complex formation on the functionality and toxic activity of Hbl. Thus it can be elucidated if the complexes either support or hinder pore formation and if—next to Nhe—another example can be found for the so far unique, permeable "pro-pores", or if Hbl conceals a different mechanism of hetero-oligomerization leading to pore formation.

4. Materials and Methods

4.1. Ethics Statement

Immunizations of mice for generating monoclonal antibodies were conducted in compliance with the German Law for Protection of Animals. Study permission was obtained by the Government of Upper Bavaria (permit number 55.2-1-54-2532.6-2-12).

4.2. Cell Lines and Culture Conditions

Vero cells were obtained from ECACC (European Collection of Cell Cultures) and cultured in 80 cm^2 culture flasks in a humidified incubator at 37 °C and 7% CO_2 in medium recommended by the supplier.

4.3. Bacterial Strains and Culture Conditions

In this study, the *B. cereus* strains F837/76 (DSM 4222), B4ac (DSM 4384), MHI 1532 (producing high amounts of Hbl toxin) and F837/76 Δ*nheABC* (producing Hbl, but no Nhe toxin) were used. For collection of toxin-rich supernatants, cells were grown in CGY medium with 1% glucose and treated as previously described [33]. Recombinant Hbl proteins were overexpressed in the *E. coli* strain BL21 (DE3), which was grown in LB medium containing 100 μg/mL ampicillin.

4.4. Construction of the nheABC Deletion Mutant

The *nheABC* deletion mutant was constructed as described previously with minor modifications [44]. Briefly, for the in-frame deletion of the *nheABC* gene cluster, an 808 bp upstream fragment of bcf_09260 (*nheA*) and an 813 bp downstream fragment of bcf_09270 (*nheC*) were amplified by PCR using the primer pairs nheA-BamHI/nheA-Spc (CG**GGATCC**CGAGTTACTGTCGTTATACC; CGTTAGCGTTTAAGTACATCCCCTGTAATTAAAGTCTTTTTCAC) and nheC-Spc/nheC-EcoRI (GCGTCCTCTTGTGAAATTAGAGGATTATACAGAAAAATTACATGAAG; CG**GAATTC**CTCCA TTATACGGTTCACTCG). To allow the selection of positive clones, the spectinomycin-resistance cassette from the TOPO/Spc plasmid was amplified with oligonucleotides For_Spc-K/Rev_Spc-K (GGATGTACTTAAACGCTAACG/CTCTAATTTCACAAGAGGACGC) and inserted between upstream and downstream fragments using ligation-independent cloning (LIC) of PCR products [45].

Following the next PCR using the primer pair nheA-BamHI/nheC-EcoRI and the LIC construct as a template, the resulting fragment was cloned into the multiple cloning site of the conjugative suicide vector pAT113 via BamHI and EcoRI, giving rise to pAT113-SpcΔnheABC. *E. coli* JM83/pRK24 was transformed with pAT113-SpcΔnheABC and the resulting strain was used for transconjugal transfer into *B. cereus* F837/76. Conjugation was carried out as described [46]. Transconjugants were screened for spectinomycin resistance and erythromycin sensitivity. Gene cluster deletion and integration of the resistance cassette resulting from a double-crossover recombination event were confirmed by PCR and sequencing.

4.5. Cloning of Recombinant Hbl (rHbl) Components

The genes encoding Hbl L$_2$ (*hblC*; bcf_15295), Hbl L$_1$ (*hblD*; bcf_15290) and Hbl B (*hblA*; bcf_15285) were amplified using the primer pairs HblL$_2$-fw-KpnI (ATAT<u>GGTACC</u>CCAAGCAGAA ACTCAACAAGAAA) and HblL$_2$-rev-NcoI (ATAT<u>CCATGG</u>TCAAAATTT ATACACTTGTTCTTC), HblL$_1$-fw-SacII (ATAT<u>CCGCGG</u>ATGGCACAAGAAACGACCGCTCAAG) and HblL$_1$-rev-NcoI (ATAT <u>CCATGG</u>GCCTCCTGTTTAAAAGCAATATC) and HblB-fw-KpnI (ATAT<u>GGTACC</u>CGCAAGTGA AATTGAACAAACGAAC) and HblB-rev-NcoI (ATAT<u>CCATGG</u>CTATTTTTGTGGAGTAACAGTT TCTAC), respectively. Chromosomal DNA from strain F837/76 was used as template. With the chosen primers, genes were amplified lacking the sequences for the N-terminal signal peptides for secretion [47,48]. *hblD* was cloned into pASK-IBA3plus (iba lifesciences), adding a C-terminal strep-tag to the recombinant Hbl L$_1$ protein. *hblC* and *hblA* were cloned into pASK-IBA5plus (iba lifesciences), resulting in addition of N-terminal strep-tags. Sequencing was performed using the primers pASK-IBA-seq-fw (CACTCCCTATCAGTGATAG) and pASK-IBA-seq-rev (GCACAATGTGCGCCAT).

4.6. Overexpression of rHbl Components

For overexpression of the rHbl components the *E. coli* strain BL21 (DE3) was used. Fifty millimetres LB medium with 100 µg/mL ampicillin were inoculated to an OD$_{600}$ of 0.15. Cells were grown at 37 °C to an OD$_{600}$ of 0.8–0.9. After overexpression was induced by adding 1 µg/mL doxycycline, cells were grown for further four hours and then harvested by centrifugation.

4.7. Purification of rHbl Components

The *E. coli* cell pellet was dissolved in 15 mL resuspension buffer (100 mM Tris, 500 mM NaCl, 1 mM EDTA, pH 8) containing protease inhibitor complete (Roche) and 100 µg/mL lysozyme. Cells were subsequently disrupted in an ultrasonic bath for 2 × 15 min on ice. Cell debris was removed by centrifugation for 30 min at 14,000× *g* and 4 °C. The remaining cell extract was filtered through a 0.2 nm filter. All strep-tagged Hbl components were purified via a 0.5 mL Strep-Tactin®-Sepharose® matrix (iba lifesciences) in a Poly-Prep chromatography column (Biorad) according to the protocol provided by the supplier. Proteins were eluted in buffer containing 100 mM Tris, 500 mM NaCl, 1 mM EDTA and 2.5 mM D-desthiobiotin, pH 8. Protein concentrations were determined after SDS-PAGE by staining with Sypro Ruby in comparison to a BSA concentration standard.

4.8. Production of Monoclonal Antibodies (mAbs)

For the preparation of the immunogen, toxin was purified via immunoaffinity chromatography (IAC) from a total of 300 mL supernatant of *B. cereus* strain MHI 1532 grown in CGY medium. For this purpose, the previously established Hbl B-specific mAb 1B8 (10 mg) [29] was coupled to 1 g CNBr-activated sepharose 4B. The column was washed with 20 mL PBS before the toxin-containing culture supernatant was applied. After a washing step with 20 mL PBS, the toxin was eluted by addition of 16 mL glycine/HCl solution (pH 2.5). Subsequently, the eluate was neutralized with 1 M Tris (pH 7.0) and dialyzed three times overnight at 4 °C against PBS. Five 12-week-old female mice (BALB/c and BALB/c × [NZW × NZB]) were immunized with this preparation. Each mouse received 15 µg purified toxin (immunization and first booster injection after 9 weeks) and another 12 µg (second booster injection after 15 weeks) emulsified in Sigma adjuvant. Three days before cell fusion each mouse received a final injection of 20 µg toxin diluted in 0.9% NaCl-solution. Cell fusion experiments, establishment of hybridomas, mass production and antibody purification were performed as described previously [20,32].

4.9. SDS-PAGE, Sypro Staining and Immunoblotting

SDS-PAGE analyses were carried out on a PhastGel gradient (10 to 15%) minigel system (GE Healthcare, Munich, Germany). For Sypro staining, proteins were fixed on the gel for 2×30 min in 50% MeOH and 7% acetic acid. The gel was then incubated with 2 mL Sypro Ruby protein stain overnight at room temperature. After 30 min washing in 10% MeOH and 7% acetic acid and additional washing for 10 min in H2O, fluorescence signals were detected on a Kodak imager (Eastman Kodak Company, Rochester, NY, USA).

For immunoblotting, proteins were blotted to a PVDF-P membrane (Millipore, Billerica, MA, USA), blocked in 3% casein-PBS and incubated with 2 µg/mL mAbs 8B12 (Hbl L2) [32], 1E9 (Hbl L1) [29], 1B8 (Hbl B) [29], 1E11 or 2B11 (NheB) [20], 1A8 (NheA) [20], 3D6 (NheC) [21], mAb 1H9 (this study), or the strep-specific StrepMAB-Classic (iba lifesciences) for 1 h at room temperature. After 3 washing steps in PBS with 0.1% Tween 20, a 1:2000 dilution of rabbit anti-mouse-horseradish peroxidase conjugate (Dako) was used as secondary antibody. After three further washing steps in PBS with 0.1% Tween 20 and two in PBS, Super Signal Western Femto Maximum Sensitivity Substrate (Pierce) was applied. Chemiluminescence signals were detected on a Kodak imager (Eastman Kodak).

4.10. Labelling of mAbs

For use as detection antibodies in the sandwich EIAs, mAbs were coupled to activated peroxidase (HRP) according to the instructions of the manufacturer (Roche). The resulting conjugate was stabilized with 1% BSA and StabilZyme® HRP Conjugate Stabilizer (SurModics) and conserved with 0.01% Thimerosal.

For flow cytometric analyses, mAb 1G8 was labelled with Alexa Fluor® 488 (Thermo Scientific, Waltham, MA, USA). For that, 2 mg 1G8 (in 1 mL PBS) were mixed with 200 µL Alexa Fluor® 488. The mixture was stirred gently for 1 h at room temperature in the dark. To remove unbound Alexa dye, 10 mL PBS were added and the material was transferred to a Amicon® Ultra-4 centrifugal filter unit (Millipore) and centrifuged for 20 min (3000 rpm, 4 °C). The labelled mAb was washed with another 10 mL PBS by repeating the centrifugation step, and afterwards resuspended in PBS. NaN$_3$ (0.1%) was added for conservation and 1% BSA for stabilization.

4.11. Flow Cytometry

For flow cytometry analyses, Vero cells were adjusted to 1×10^6 cells in 1 mL EC buffer (140 mM NaCl, 15 mM HEPES, 1 mM MgCl$_2$, 1 mM CaCl$_2$, 10 mM glucose pH 7.2). Then 6.5 pmol/mL rHbl B and, if applicable, mAbs were added (ratio rHbl B:mAb = 1:1). The mixture was incubated for 1 h at 37 °C under moderate agitation. Then, 2 mL 1% BSA-PBS were added and cells were centrifuged for 5 min at 800 rpm. Cells were washed again in 2 mL 1% BSA-PBS. When rHbl B and mAb 1D7 were applied successively, the incubation step at 37 °C and the washing step were repeated. For detection of cell-bound rHbl B, the samples were incubated with 3 µg/mL Hbl B-specific Alexa Fluor® 488-labelled mAb 1G8 for 1 h at 4 °C. After additional washing, samples were resuspended in 500 µL 1% BSA-PBS and subsequently analysed in a FACS Calibur using the CellQuestPro software (BD Bioscience).

4.12. Enzyme Immunoassays (EIAs)

Indirect and sandwich enzyme immunoassays were performed as described before [20,28,32,49]. For detection of rHbl components, established mAbs were used (1A12 and 8B12 [32] for Hbl L$_2$, 1E9 [29] for Hbl L$_1$ and 1B8 [29] for Hbl B), as well as the ones newly generated in this study (Table S1). To determine relative affinities of the newly generated mAbs, *B. cereus* culture supernatants and rHbl components were used in indirect EIAs. For that, dilution series of the antigens were applied, followed by the cell culture supernatants (constantly 1:20 in PBS). The relative affinity corresponds with the dilution that results in an absorbance value of 1.0. The productivity of the hybridoma cell lines was also determined in indirect EIAs. Supernatant of *B. cereus* MHI 1532 was applied (constantly 1:10 in PBS), followed by dilution series of the cell culture supernatants.

rHbl complex formation was investigated using indirect and sandwich EIAs. In the indirect assay, the microtiter plate was coated with a serial dilution of rHbl L_1 (120–0 pmol/mL) overnight. After washing, the second rHbl component was applied in constant concentration (60 pmol/mL) for 1 h. After blocking for 30 min with 3% sodium-caseinate in PBS, HblB-specific mAb 1B8 [29] and Hbl L_2-specific mAb 1H9 (this study) (2 µg/mL in PBS) were applied, respectively. After additional washing, rabbit-anti-mouse-HRP conjugate (1:2000 in 1% sodium-caseinate in PBS) was applied for detection. In the sandwich assay, the microtiter plate was coated with 10 µg/mL mAbs 1E9 [29] or 1G8 (this study), both Hbl L_1-specific, overnight. While the plate was blocked for 30 min with 3% sodium-caseinate in PBS, a mixture of rHbl components (L_1+B or L_1+L_2; 1.5 pmol/µL each, ratios 1:1, 5:1, 1:5, 10:1 or 1:10) was incubated at RT. The mixtures were applied to the microtiter plate as serial dilution from 75 to 0 pmol/mL. After washing, a specific mAb conjugate was applied for detection (1:2000 in 1% sodium-caseinate in PBS; 1B8-HRP [29] against Hbl B or 1H9-HRP (this study) against Hbl L_2). Analogously to the detection of the rHbl components, sandwich EIAs were used to detect Hbl complexes in the supernatant of *B. cereus* strain F837/76, which was applied to the microtiter plates as serial dilution. One-site-binding curves were applied to depict the absorption values compared to the sample dilutions.

4.13. Dot Blot

One hundred microlitres per dot of the first purified rHbl component, diluted in PBS, were applied to a PVDF membrane (Immobilon-P, Millipore, USA) as dilution series from 480 to 3.75 pmol. The membrane was removed from the dotting chamber and blocked with 3% sodium-caseinate-PBS overnight. Subsequently, the second purified rHbl component was applied (30 pmol in PBS) and overlaid on the dots for 1 h. The membrane was washed for 3 × 10 min in PBS containing 0.1% Tween 20. After that, Hbl-specific mAbs were applied (1B8 against Hbl B [29], 1H9 against Hbl L_2 (this study); 1E9 [29] and 1G8 (this study) against Hbl L_1; 3 µg/mL in 3% sodium-caseinate-PBS with 0.025% Tween 20) and overlaid on the dots for 1 h. The membrane was again washed for 3 × 10 min in PBS containing 0.1% Tween 20 before anti-mouse-IgG-alkaline-phosphatase-conjugate (Sigma) was applied (1:10.000 in 3% sodium-caseinate-PBS with 0.025% Tween 20) for 1 h. Again, the membrane was washed for 3 × 10 min in PBS containing 0.1% Tween 20 and for 2 × 10 min in PBS before signals were detected using NBT/BCIP solution (Roche).

4.14. Surface Plasmon Resonance (SPR)

Binding measurements were performed on a BIACORE 3000 instrument (Biacore Inc., Piscataway, NJ, USA). In a first experiment, rHbl L_1 was coupled and rHbl L_2 and B were used as ligands. For that, L_1 was diluted to a final concentration of 416 nM in 10 mM sodium acetate, pH 4.8, and chemically immobilized (amine coupling, 560 RU bound) onto a CM5 sensor chip (Biacore Inc.). The L_2 and B protein samples were diluted in running buffer (PBS, 1 mM DTT and 0.005% Tween 20) and injected over the sensor chip surface at 30 µL/min at 20 °C. The samples were injected onto the sensor chip from the lowest to the highest concentration with injection of 250 nM ligand in duplicate within each experiment. L_2 and B protein samples were measured three and four times, respectively.

For the second experiment, rHbl B was diluted to a final concentration of 350 nM in 10 mM sodium acetate, pH 4.0. 500 RU were coupled onto a CM5 sensor chip. rHbl L_2 was diluted in the same running buffer as above and injected over the sensor chip surface at 30 µL/min at 20 °C. The experiment was repeated four times. Background subtraction was done using an unmodified sensor chip surface. Data were analysed using BIAevaluation program (Biacore Inc.). For each measurement the equilibrium dissociation constant (K_D) was calculated from steady-state measurement. The K_{DS} from three or four experiments (see above) were used to calculate mean values and standard deviations. Despite long injection times the measurement of $L_1 + B$ did not reach full binding saturation for lower protein concentrations. Therefore, the derived K_D value is likely to be slightly underestimated.

4.15. Cytotoxicity Assays

Stock solutions (1.5 pmol/µL) of the rHbl components were used for toxicity assays on Vero cells. They were pre-mixed in appropriate ratios and added in 1:40 dilutions to the cells. Propidium iodide (PI) influx tests were performed as described before [33,49]. WST-1 bioassays were carried out as previously described [20,32,49]. For neutralization assays, 10 µg/well purified mAbs were applied constantly with a serial dilution (starting 1:20) of culture supernatant of *B. cereus* strain F837/76 Δ*nheABC*.

Supplementary Materials: The following are available online at www.mdpi.com/2072-6651/9/9/288/s1. **Table S1:** Characteristics of the established hybridoma cell lines and mAbs, **Table S2:** Statistics of the curves fitted to the EIA and SPR experiments, **Figure S1:** Detection of Hbl complex formation via Dot blot.

Acknowledgments: This work was supported by the Federal Ministry of Education and Research (BMBF) of Germany (Food supply and analysis LEVERA, funding code 13N12611).

Author Contributions: F.T. prepared immunogens, characterized mAbs and complexes and wrote parts of the manuscript. R.D. immunized mice, generated hybridoma cell lines and was involved in experimental setup and writing of the manuscript. K.S. generated the F837/76 Δ*nheABC* mutant. R.J. performed SPR measurements. D.N. and E.M. were involved in experimental setup and writing of the manuscript. N.J. cloned Hbl genes and purified the corresponding proteins, performed hybrid sandwich EIAs, characterized the complexes and wrote the manuscript.

Conflicts of Interest: The authors declare no conflict of interest. The founding sponsors had no role in the design of the study; in the collection, analyses, or interpretation of data; in the writing of the manuscript, and in the decision to publish the results.

References

1. Stenfors Arnesen, L.P.; Fagerlund, A.; Granum, P.E. From soil to gut: *Bacillus cereus* and its food poisoning toxins. *FEMS Microbiol. Rev.* **2008**, *32*, 579–606. [CrossRef] [PubMed]

2. Andersson, A.; Ronner, U.; Granum, P.E. What problems does the food industry have with the spore-forming pathogens *Bacillus cereus* and *Clostridium perfringens*? *Int. J. Food Microbiol.* **1995**, *28*, 145–155. [CrossRef]

3. Ehling-Schulz, M.; Frenzel, E.; Gohar, M. Food-bacteria interplay: Pathometabolism of emetic *Bacillus cereus*. *Front. Microbiol.* **2015**, *6*, 704. [CrossRef] [PubMed]

4. Lund, T.; De Buyser, M.L.; Granum, P.E. A new cytotoxin from *Bacillus cereus* that may cause necrotic enteritis. *Mol. Microbiol.* **2000**, *38*, 254–261. [CrossRef] [PubMed]

5. Dierick, K.; Van Coillie, E.; Swiecicka, I.; Meyfroidt, G.; Devlieger, H.; Meulemans, A.; Hoedemaekers, G.; Fourie, L.; Heyndrickx, M.; Mahillon, J. Fatal family outbreak of *Bacillus cereus*-associated food poisoning. *J. Clin. Microbiol.* **2005**, *43*, 4277–42799. [CrossRef] [PubMed]

6. Naranjo, M.; Denayer, S.; Botteldoorn, N.; Delbrassinne, L.; Veys, J.; Waegenaere, J.; Sirtaine, N.; Driesen, R.B.; Sipido, K.R.; Mahillon, J.; et al. Sudden death of a young adult associated with *Bacillus cereus* food poisoning. *J. Clin. Microbiol.* **2011**, *49*, 4379–4381. [CrossRef] [PubMed]

7. Agata, N.; Ohta, M.; Mori, M.; Isobe, M.A. A novel dodecadepsipeptide, cereulide, is an emetic toxin of *Bacillus cereus*. *FEMS Microbiol. Lett.* **1995**, *129*, 17–20. [CrossRef] [PubMed]

8. Ehling-Schulz, M.; Fricker, M.; Scherer, S. *Bacillus cereus*, the causative agent of an emetic type of food-borne illness. *Mol. Nutr. Food Res.* **2004**, *48*, 479–487. [CrossRef] [PubMed]

9. Clavel, T.; Carlin, F.; Lairon, D.; Nguyen-The, C.; Schmitt, P. Survival of *Bacillus cereus* spores and vegetative cells in acid media simulating human stomach. *J. Appl. Microbiol.* **2004**, *97*, 214–219. [CrossRef] [PubMed]

10. Beecher, D.J.; Schoeni, J.L.; Wong, A.C. Enterotoxic activity of hemolysin BL from *Bacillus cereus*. *Infect. Immun.* **1995**, *63*, 4423–4428. [PubMed]

11. Lund, T.; Granum, P.E. Characterisation of a non-haemolytic enterotoxin complex from *Bacillus cereus* isolated after a foodborne outbreak. *FEMS Microbiol. Lett.* **1996**, *141*, 151–156. [CrossRef] [PubMed]

12. Hardy, S.P.; Lund, T.; Granum, P.E. CytK toxin of *Bacillus cereus* forms pores in planar lipid bilayers and is cytotoxic to intestinal epithelia. *FEMS Microbiol. Lett.* **2001**, *197*, 47–51. [CrossRef] [PubMed]

13. Fagerlund, A.; Ween, O.; Lund, T.; Hardy, S.P.; Granum, P.E. Genetic and functional analysis of the *cytK* family of genes in *Bacillus cereus*. *Microbiology* **2004**, *150*, 2689–2697. [CrossRef] [PubMed]

14. Guinebretiére, M.H.; Auger, S.; Galleron, N.; Contzen, M.; De Sarrau, B.; De Buyser, M.L.; Lamberet, G.; Fagerlund, A.; Granum, P.E.; Lereclus, D.; et al. *Bacillus cytotoxicus* sp. nov. is a novel thermotolerant species of the *Bacillus cereus* Group occasionally associated with food poisoning. *Int. J. Syst. Evol. Microbiol.* **2013**, *63*, 31–40. [CrossRef] [PubMed]

15. Lindbäck, T.; Fagerlund, A.; Rødland, M.S.; Granum, P.E. Characterization of the *Bacillus cereus* Nhe enterotoxin. *Microbiology* **2004**, *150*, 3959–3967. [CrossRef] [PubMed]

16. Fagerlund, A.; Lindbäck, T.; Storset, A.K.; Granum, P.E.; Hardy, S.P. *Bacillus cereus* Nhe is a pore-forming toxin with structural and functional properties similar to the ClyA (HlyE, SheA) family of haemolysins, able to induce osmotic lysis in epithelia. *Microbiology* **2008**, *154*, 693–704. [CrossRef] [PubMed]

17. Lindbäck, T.; Hardy, S.P.; Dietrich, R.; Sødring, M.; Didier, A.; Moravek, M.; Fagerlund, A.; Bock, S.; Nielsen, C.; Casteel, M.; et al. Cytotoxicity of the *Bacillus cereus* Nhe enterotoxin requires specific binding order of its three exoprotein components. *Infect. Immun.* **2010**, *78*, 3813–3821. [CrossRef] [PubMed]

18. Didier, A.; Dietrich, R.; Gruber, S.; Bock, S.; Moravek, M.; Nakamura, T.; Lindbäck, T.; Granum, P.E.; Märtlbauer, E. Monoclonal antibodies neutralize *Bacillus cereus* Nhe enterotoxin by inhibiting ordered binding of its three exoprotein components. *Infect. Immun.* **2012**, *80*, 832–838. [CrossRef] [PubMed]

19. Sastalla, I.; Fattah, R.; Coppage, N.; Nandy, P.; Crown, D.; Pomerantsev, A.P.; Leppla, S.H. The *Bacillus cereus* Hbl and Nhe tripartite enterotoxin components assemble sequentially on the surface of target cells and are not interchangeable. *PLoS ONE* **2013**, *8*, e76955. [CrossRef] [PubMed]

20. Dietrich, R.; Moravek, M.; Bürk, C.; Granum, P.E.; Märtlbauer, E. Production and characterization of antibodies against each of the three subunits of the *Bacillus cereus* nonhemolytic enterotoxin complex. *Appl. Environ. Microbiol.* **2005**, *71*, 8214–8220. [CrossRef] [PubMed]

21. Heilkenbrinker, U.; Dietrich, R.; Didier, A.; Zhu, K.; Lindbäck, T.; Granum, P.E.; Märtlbauer, E. Complex formation between NheB and NheC is necessary to induce cytotoxic activity by the three-component *Bacillus cereus* Nhe enterotoxin. *PLoS ONE* **2013**, *8*, e63104. [CrossRef] [PubMed]

22. Beecher, D.J.; MacMillan, J.D. A novel bicomponent hemolysin from *Bacillus cereus*. *Infect. Immun.* **1990**, *58*, 2220–2227. [PubMed]

23. Beecher, D.J.; Macmillan, J.D. Characterization of the components of hemolysin BL from *Bacillus cereus*. *Infect. Immun.* **1991**, *59*, 1778–1784. [PubMed]

24. Beecher, D.J.; Wong, A.C. Tripartite hemolysin BL from *Bacillus cereus*. Hemolytic analysis of component interactions and a model for its characteristic paradoxical zone phenomenon. *J. Biol. Chem.* **1997**, *272*, 233–239. [CrossRef] [PubMed]

25. Beecher, D.J.; Wong, A.C. Tripartite haemolysin BL: Isolation and characterization of two distinct homologous sets of components from a single *Bacillus cereus* isolate. *Microbiology* **2000**, *146*, 1371–1380. [CrossRef] [PubMed]

26. Madegowda, M.; Eswaramoorthy, S.; Burley, S.K.; Swaminathan, S. X-ray crystal structure of the B component of Hemolysin BL from *Bacillus cereus*. *Proteins* **2008**, *71*, 534–540. [CrossRef] [PubMed]

27. Guinebretiére, M.H.; Broussolle, V.; Nguyen-The, C. Enterotoxigenic profiles of food-poisoning and food-borne *Bacillus cereus* strains. *J. Clin. Microbiol.* **2002**, *40*, 3053–3056. [CrossRef] [PubMed]

28. Moravek, M.; Dietrich, R.; Bürk, C.; Broussolle, V.; Guinebretiére, M.H.; Granum, P.E.; Nguyen-The, C.; Märtlbauer, E. Determination of the toxic potential of *Bacillus cereus* isolates by quantitative enterotoxin analyses. *FEMS Microbiol. Lett.* **2006**, *257*, 293–298. [CrossRef] [PubMed]

29. Wehrle, E.; Moravek, M.; Dietrich, R.; Bürk, C.; Didier, A.; Märtlbauer, E. Comparison of multiplex PCR, enzyme immunoassay and cell culture methods for the detection of enterotoxinogenic *Bacillus cereus*. *J. Microbiol. Methods* **2009**, *78*, 265–270. [CrossRef] [PubMed]

30. Beecher, D.J.; Wong, A.C. Improved purification and characterization of hemolysin BL, a hemolytic dermonecrotic vascular permeability factor from *Bacillus cereus*. *Infect. Immun.* **1994**, *62*, 980–986. [PubMed]

31. Ryan, P.A.; Macmillan, J.D.; Zilinskas, B.A. Molecular cloning and characterization of the genes encoding the L1 and L2 components of hemolysin BL from *Bacillus cereus*. *J. Bacteriol.* **1997**, *179*, 2551–2556. [CrossRef] [PubMed]

32. Dietrich, R.; Fella, C.; Strich, S.; Märtlbauer, E. Production and characterization of monoclonal antibodies against the hemolysin BL enterotoxin complex produced by *Bacillus cereus*. *Appl. Environ. Microbiol.* **1999**, *65*, 4470–4474. [PubMed]

33. Jessberger, N.; Dietrich, R.; Bock, S.; Didier, A.; Märtlbauer, E. *Bacillus cereus* enterotoxins act as major virulence factors and exhibit distinct cytotoxicity to different human cell lines. *Toxicon* **2014**, *77*, 49–57. [CrossRef] [PubMed]

34. Zhu, K.; Didier, A.; Dietrich, R.; Heilkenbrinker, U.; Waltenberger, E.; Jessberger, N.; Märtlbauer, E.; Benz, R. Formation of small transmembrane pores: An intermediate stage on the way to *Bacillus cereus* non-hemolytic enterotoxin (Nhe) full pores in the absence of NheA. *Biochem. Biophys. Res. Commun.* **2016**, *469*, 613–618. [CrossRef] [PubMed]

35. Granum, P.E.; O'Sullivan, K.; Lund, T. The sequence of the non-haemolytic enterotoxin operon from *Bacillus cereus*. *FEMS Microbiol. Lett.* **1999**, *177*, 225–229. [CrossRef] [PubMed]

36. Didier, A.; Dietrich, R.; Märtlbauer, E. Antibody Binding Studies Reveal Conformational Flexibility of the *Bacillus cereus* Non-Hemolytic Enterotoxin (Nhe) A-Component. *PLoS ONE* **2016**, *11*, e0165135. [CrossRef] [PubMed]

37. Tirado, S.M.; Yoon, K.J. Antibody-dependent enhancement of virus infection and disease. *Viral Immunol.* **2003**, *16*, 69–86. [CrossRef] [PubMed]

38. Takada, A.; Kawaoka, Y. Antibody-dependent enhancement of viral infection: Molecular mechanisms and in vivo implications. *Rev. Med. Virol.* **2003**, *13*, 387–398. [CrossRef] [PubMed]

39. Guzman, M.G.; Alvarez, M.; Halstead, S.B. Secondary infection as a risk factor for dengue hemorrhagic fever/dengue shock syndrome: An historical perspective and role of antibody-dependent enhancement of infection. *Arch. Virol.* **2013**, *158*, 1445–1459. [CrossRef] [PubMed]

40. Willey, S.; Aasa-Chapman, M.; O'Farrell, S.; Pellegrino, P.; Williams, I.; Weiss, R.A.; Neil, S.J.D. Extensive complement-dependent enhancement of HIV-1 by autologous non-neutralising antibodies at early stages of infection. *Retrovirology* **2011**, *8*, 16. [CrossRef] [PubMed]

41. Takano, T.; Kawakami, C.; Yamada, S.; Satoh, R.; Hohdatsu, T. Antibody-dependent enhancement occurs upon re-infection with the identical serotype virus in feline infectious peritonitis virus infection. *J. Vet. Med. Sci.* **2008**, *70*, 1315–1321. [CrossRef] [PubMed]

42. Takada, A.; Feldmann, H.; Ksiazek, T.G.; Kawaoka, Y. Antibody-dependent enhancement of Ebola virus infection. *J. Virol.* **2003**, *77*, 7539–7544. [CrossRef] [PubMed]

43. Mohamed, N.; Li, J.; Ferreira, C.S.; Little, S.F.; Friedlander, A.M.; Spitalny, G.L.; Casey, L.S. Enhancement of anthrax lethal toxin cytotoxicity: A subset of monoclonal antibodies against protective antigen increases lethal toxin-mediated killing of murine macrophages. *Infect. Immun.* **2004**, *72*, 3276–3283. [CrossRef] [PubMed]

44. Lücking, G.; Dommel, M.K.; Scherer, S.; Fouet, A.; Ehling-Schulz, M. Cereulide synthesis in emetic *Bacillus cereus* is controlled by the transition state regulator AbrB, but not by the virulence regulator PlcR. *Microbiology* **2009**, *155*, 922–931. [CrossRef] [PubMed]

45. Aslanidis, C.; de Jong, P.J. Ligation-independent cloning of PCR products (LIC-PCR). *Nucleic Acids Res.* **1990**, *18*, 6069–6074. [CrossRef] [PubMed]

46. Pezard, C.; Berche, P.; Mock, M. Contribution of individual toxin components to virulence of *Bacillus anthracis*. *Infect. Immun.* **1991**, *59*, 3472–3477. [PubMed]

47. Fagerlund, A.; Lindbäck, T.; Granum, P.E. *Bacillus cereus* cytotoxins Hbl, Nhe and CytK are secreted via the Sec translocation pathway. *BMC Microbiol.* **2010**, *10*, 304. [CrossRef] [PubMed]

48. Økstad, O.A.; Gominet, M.; Purnelle, B.; Rose, M.; Lereclus, D.; Kolstø, A.B. Sequence analysis of three *Bacillus cereus* loci carrying PlcR-regulated genes encoding degradative enzymes and enterotoxin. *Microbiology* **1999**, *145*, 3129–3138. [CrossRef] [PubMed]

49. Jessberger, N.; Rademacher, C.; Krey, V.M.; Dietrich, R.; Mohr, A.K.; Böhm, M.E.; Scherer, S.; Ehling-Schulz, M.; Märtlbauer, E. Simulating Intestinal Growth Conditions Enhances Toxin Production of Enteropathogenic *Bacillus cereus*. *Front. Microbiol.* **2017**, *8*, 627. [CrossRef] [PubMed]

Article

The Vip3Ag4 Insecticidal Protoxin from *Bacillus thuringiensis* Adopts A Tetrameric Configuration That Is Maintained on Proteolysis

Leopoldo Palma [1,†], David J. Scott [2,3,4], Gemma Harris [3], Salah-Ud Din [5,‡], Thomas L. Williams [6], Oliver J. Roberts [5], Mark T. Young [5], Primitivo Caballero [1] and Colin Berry [5,*]

1   Instituto de Agrobiotecnología, CSIC-UPNA-Gobierno de Navarra, Campus Arrosadía, Mutilva 31192, Navarra, Spain; lpalma.leopoldo@gmail.com (L.P.); pcm92@unavarra.es (P.C.)
2   School of Biosciences, University of Nottingham, Sutton Bonnington Campus, Leicestershire LE12 5RD, UK; david.scott@nottingham.ac.uk
3   Research Complex at Harwell, Rutherford Appleton Laboratory, Harwell Campus, Oxfordshire OX11 0FA, UK; gemma.harris@rc-harwell.ac.uk
4   ISIS Spallation Neutron and Muon Source, Rutherford Appleton Laboratory, Harwell Campus, Oxfordshire OX11 0QX, UK
5   Cardiff School of Biosciences, Cardiff University, Park Place, Cardiff CF10 3AT, UK; salahuddin@cemb.edu.pk (S.-U.D.); o.roberts1@nhs.net (O.J.R.); youngmt@cardiff.ac.uk (M.T.Y.)
6   Cardiff School of Chemistry, Cardiff University, Park Place, Cardiff CF10 3AT, UK; williamst30@cardiff.ac.uk
*   Correspondence: berry@cardiff.ac.uk; Tel.: +44-29-2087-4508
†   Current address: Centro de Investigaciones y Transferencia de Villa María (CITVM-CONICET), Universidad Nacional de Villa María, Villa María, Córdoba 5900, Argentina.
‡   Current address: National Center of Excellence in Molecular Biology (CEMB), University of the Punjab, Lahore 54590, Pakistan.

Academic Editor: Shin-ichi Miyoshi
Received: 17 January 2017; Accepted: 12 May 2017; Published: 14 May 2017

Abstract: The Vip3 proteins produced during vegetative growth by strains of the bacterium *Bacillus thuringiensis* show insecticidal activity against lepidopteran insects with a mechanism of action that may involve pore formation and apoptosis. These proteins are promising supplements to our arsenal of insecticidal proteins, but the molecular details of their activity are not understood. As a first step in the structural characterisation of these proteins, we have analysed their secondary structure and resolved the surface topology of a tetrameric complex of the Vip3Ag4 protein by transmission electron microscopy. Sites sensitive to proteolysis by trypsin are identified and the trypsin-cleaved protein appears to retain a similar structure as an octomeric complex comprising four copies each of the ~65 kDa and ~21 kDa products of proteolysis. This processed form of the toxin may represent the active toxin. The quality and monodispersity of the protein produced in this study make Vip3Ag4 a candidate for more detailed structural analysis using cryo-electron microscopy.

Keywords: Vip3 toxin; electron microscopy; surface topology

---

1. Introduction

Bacillus thuringiensis (Bt) is a Gram-positive entomopathogenic bacterium with strains that show toxicity to a wide variety of invertebrates [1]. The best-studied toxins produced by these strains are the crystal (Cry) and cytolytic (Cyt) toxins, also known as δ-endotoxins, which are synthesized during the stationary growth phase and into sporulation as parasporal crystalline inclusions [2]. In addition, Bt synthesizes other insecticidal proteins that are secreted to the culture medium during the vegetative growth phase: vegetative insecticidal proteins (Vip) [3,4] and secreted insecticidal proteins (Sip) [5].

Vip proteins have been classified into four families; Vip1, Vip2, Vip3 and Vip4, according to their degree of amino acid similarity [6]. Vip1 and Vip2 act together as a binary toxin with insecticidal activity against some coleopteran [4] and hemipteran pests [7] and function through the ADP-ribosyltransferase activity of Vip2 [8], the structure of which has been solved [9]. The Vip4 protein is distantly related to the Vip1 component of this binary toxin, but its activity remains unknown to date [10].

Vip3 proteins have no primary sequence homology to the other Vip proteins or other toxins and exhibit toxicity against lepidopteran larvae [3,11]. As for the Cry and Cyt toxins of Bt, the Vip3 proteins are named according to the degree of amino acid identity between family members with subdivisions of the protein family having different secondary rank (denoted by the capital letter) at <78% identity, tertiary rank (denoted by the lower case letter) at <95%, and a quaternary rank (denoted by the final number) assigned to each new entry into the nomenclature [6]. Vip3Aa proteins appear to recognise distinct receptors from Cry1 toxins in *Manduca sexta* [12], *Agrotis segetum* [13] and *Spodoptera littoralis* [14], which is consistent with reports that insects resistant to Cry toxins are not cross-resistant to Vip3 proteins [12,15,16]. This has made Vip3 proteins attractive as additional insect resistance traits in transgenic crops (e.g., [17,18]).

The current understanding of the mode of action of Vip3 proteins remains limited, although a number of steps towards intoxication are known [19]. Proteolysis of the ~88 kDa full-length Vip3A proteins to ~65 kDa by trypsin or the gut juices of both susceptible and non-susceptible insects has been reported [12,13,20,21]. It has been proposed that differences in the amounts of further digestion products accumulated may be linked to levels of susceptibility to the toxins [20,21]. Binding of proteolytically processed Vip3A proteins to ligands of 55 and 100 kDa in *Ephestia kuehniella* [14], 80 and 100 kDa in *Manduca sexta* [12] or of 65 kDa in *Agrotis segetum* [13] has been reported and using a two-hybrid system, a putative ~43 kDa receptor with homology with the tenacins has been identified in *Agrotis ipsilon* [22]. Toxin activated by gut juices is able to form pores in planar lipid bilayers and dissected *Manduca sexta* gut tissue [12]. The histopathology of intoxication shows cytoplasmic vacuolization and breakup of the brush border membrane [13,14] and there is evidence that Vip3Aa causes apoptosis in *Spodoptera frugiperda* Sf9 cells [22,23]. However, an understanding of the molecular mechanism of the Vip3 proteins is hampered by the absence of structural data. As a first stage in the process of 3D-structure determination, here we describe the expression, purification and analysis of the trypsin processing of the Vip3Ag4 protein. We analyse its secondary structure and present approximately 33 Å resolution surface topologies of both a Vip3Ag4 tetramer and a trypsin-processed complex, determined via transmission electron microscopy and single particle analysis.

2. Results and Discussion

2.1. Purification of Vip3Ag4

Expression of Vip3Ag4 in *Escherichia coli* and purification by nickel affinity chromatography resulted in a single band of around 91 kDa on SDS PAGE, consistent with the expected size for the recombinant protein including His-tag (91.38 kDa). Size exclusion chromatography produced several peaks (Figure 1). The first, small peak emerging at 40 min may represent aggregated material, the largest peak (60 min) has an approximate molecular mass (calculated with respect to gel filtration standards) of 354 kDa, which corresponds approximately to a tetrameric form. There is a further shoulder to this peak (~70 min) that appears to represent monomeric Vip3Ag4. A recent study with Vip3Aa35 (82% identical to Vip3Ag4), activated with trypsin, indicated the presence of aggregated, monomeric and tetrameric forms of this protein; the proportions of these forms could be manipulated by changing buffer conditions [24]. Fractions corresponding to the putative tetrameric form of Vip3Ag4, chosen to exclude those that might include the monomeric form were combined and the protein was concentrated to 1 mg/mL. Mass spectrometric analysis of this material revealed a protein of 91,245.5 Da, a size that demonstrates the production of the monomeric, His-tagged protein, lacking the initial Met residue (theoretical mass 91,245.7). Initiator methionine residues are often removed by *E. coli* methionine aminopeptidases, especially when, as in the recombinant Vip3Ag4, the next residue has a small sidechain (in this case glycine) [25].

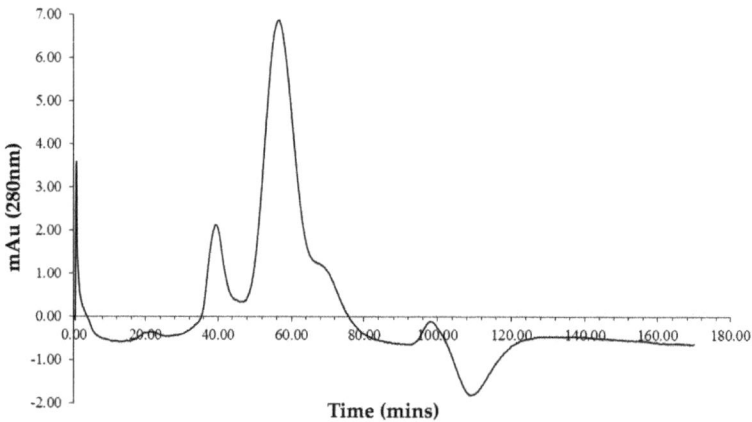

Figure 1. Size exclusion chromatography analysis of the initial Vip3Ag4 preparation. The absorption of the Vip3Ag4 at 280 nm over time is shown. The major peak, emerging around 60 min, corresponds to the Vip3 tetramer.

2.2. Trypsin Processin

Trypsin digestion of the Vip3Ag4 protein produced bands of ~65 kDa and ~21 kDa on SDS PAGE. The larger band was subjected to five cycles of N-terminal sequencing and revealed the sequence N/K-S-S-E/P-A, which is consistent with the sequence NSSPA that starts at residue N199 of the native Vip3Ag4 sequence (residue 233 of the His-tagged recombinant protein). Mass spectrometry of the products of trypsin digestion revealed two major peaks with molecular weights of 65,401.0 and 20,740.0. These masses match the region of Vip3Ag4 from N199 to the C—terminal residue (expected mass 65,401.2) and the region from D33 of the His tagged recombinant protein (two residues upstream of the initiator Met of the naturally-occurring protein) to K182 of the native Vip3Ag4 sequence (residue 216 of the recombinant protein—expected mass 20,740.7). Cleavage sites are shown on the primary sequence of the recombinant Vip3Ag4 in Figure S1. The trypsin processing indicated by these peaks would also produce a further fragment from F183 to K198 (FEDLTFATETTLKVKK) and the N-terminal region of the recombinant protein, but peaks corresponding to these fragments were not observed. The pattern of digestion seen with Vip3Ag4 is consistent with other reports of Vip3 processing by both trypsin and insect gut extracts [12,13,20,21]. Estruch et al. [22] reported a similar initial processing of Vip3Aa by gut juices from *A. ipsilon* to fragments of ~22 kDa (residues 1–198) and ~65 kDa (residues 200–789), although they did not see processing after the conserved K182 residue. Gut juice was then seen to cause further proteolysis of the ~65 kDa Vip3Aa fragment to ~45 kDa (residues 412–789) or ~33 kDa (residues 200–455). These further degradation products were not seen in the present study with Vip3Ag4. This may be due to sequence differences in the two toxins: Vip3Ag4 has the sequence KTK (residues 455–457) in place of the Vip3Aa sequence KKK, which may affect trypsin-like cleavage in this region to generate the ~33 kDa product. The presence of proteolytic activities other than trypsin in the gut juice is also likely given the cleavage between Thr411 and Asn412 that generates the ~45 kDa product [22], since this site is conserved in Vip3Ag4.

When the trypsin-treated Vip3Ag was analysed by Size exclusion chromatography (SEC), there was no significant change in the elution time for the major peak compared to unprocessed material (Figure 2a) and when fractions from this peak were analysed by SDS PAGE, both ~21 and ~65 kDa bands were seen (Figure 2b). This indicates not only that the ~21 kDa trypsin-cleaved fragment from D33-K216 of the recombinant protein remains associated with the larger ~65 kDa fragment after digestion, but also that the processed Vip3Ag toxin remains in the multimeric complex form in solution. Kunthic et al. recently described a variety of monomeric and oligomeric forms of trypsin-treated Vip3Aa35 including a form arising after the addition and dialysis of detergent that appeared to be a tetramer of the ~66 kDa polypeptide only. Our experiments suggest that, in the absence of detergent treatment, Vip3Ag4 treated with trypsin is an octameric complex consisting of tetramers of both the ~65 kDa and 21 kDa fragments. This is consistent with the finding that the trypsin cleavage products of Vip3Aa16 (82% identical to Vip3Ag4) remain associated [26] and may indicate that the proteolytically cleaved complex is the active form of the toxin. The sensitivity of *S. exigua* larvae to Vip35Aa in full-length or trypsin-processed forms shows no significant difference [24]. Since trypsin-like cleavage of the full-length form would be expected in vivo in the insect gut, these results show that trypsin cleavage does not reduce activity and the processed form may be the active agent. The stability of the 65 kDa product of Vip3Aa shows a correlation to differences in species susceptibility [20] and gut juice-processed Vip3A is able to form pores, whereas full-length protein could not [12].

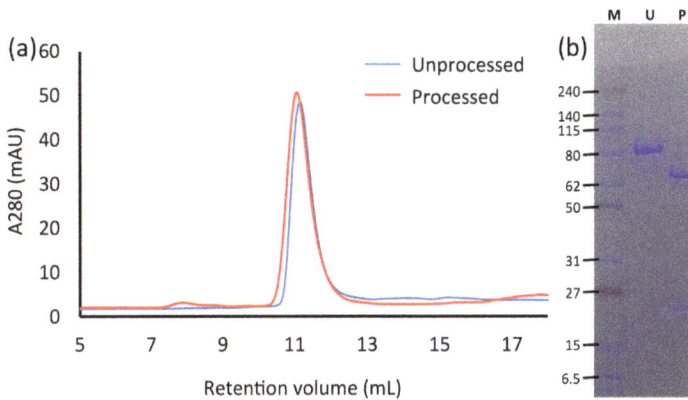

Figure 2. Analysis of purified Vip3Ag4 before and after trypsin treatment. (**a**) SEC of unprocessed and trypsin-treated recombinant Vip3Ag4; (**b**) SDS Poly Acrylamide Gel Electrophoresis of unprocessed Vip3Ag4 (U) and the SEC peak from trypsin-treated Vip3Ag4 (P); the sizes of protein markers (M) are indicated.

2.3. Circular Dichroism

To estimate the secondary structure content of the Vip3Ag4 protein, circular dichroism analysis was performed. Analysis of the data using the Dichroweb server found the best fit of data to reference datasets using the CONTIN analysis program and reference set 3 [27]. Using these tools, a good match with reference data was obtained for most of the curve (Figure S2), suggesting a secondary structure content in the tetrameric Vip3Ag4 of ~11% alpha helix, ~38% ß sheet, 22% loops and 30% unordered.

2.4. SEC-MALLS

In order to achieve an accurate reconstruction of TEM data to produce a structure, it is essential to know the multimeric form of the protein being analysed. To verify our initial observations during purification, we undertook SEC-MALLS and analytical ultracentrifugation analyses.

The elution profile of the Vip3Ag4 sample contained a single peak (Figure 3). The molecular mass obtained from the SEC-MALLS analysis was consistent with that of a Vip3Ag4 tetramer (Table S1). It was noted that the molecular mass across the peak appeared to increase from the leading- to the trailing-edge of the peak, which is suggestive of non-ideal behaviour.

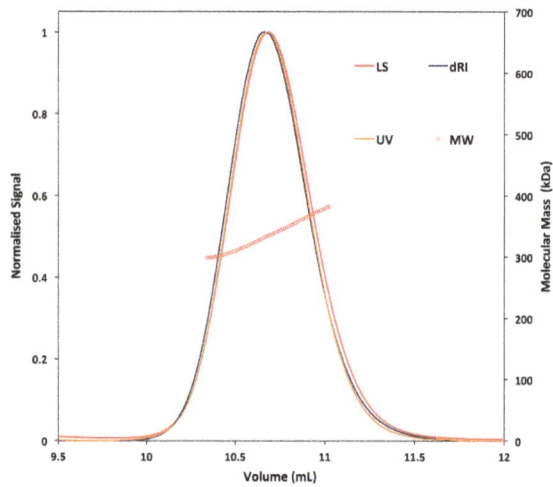

Figure 3. SEC-MALLS analysis of the oligomeric state of Vip3Ag4. LS—light scattering; dRI—differential refractive index; UV—UV at 280 nm; MW—molecular weight.

2.5. Analytical Ultracentrifugation

Analysis of Vip3Ag4 by sedimentation velocity (SV) also suggests that the protein is tetrameric (Table S2). During analysis of the interference data, corrections were made for buffer sedimentation, without which the molecular masses obtained were inconsistent with those obtained from analysis of the absorbance data and from the SEC-MALLS analysis. A decrease in sedimentation coefficient with increasing concentration was noted, suggestive of non-ideal sedimentation (Figure 4). An attempt to determine the concentration-dependent coefficient, k_s, by linear extrapolation of a plot of $1/S_{(20,w)}$ against concentration yielded very high values, suggesting that other factors apart from excluded volume, such as charge, are contributing significantly to the sedimentation behaviour of this protein.

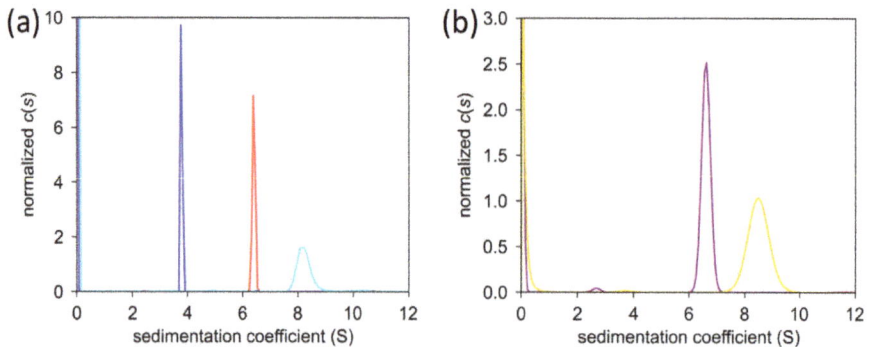

Figure 4. Sedimentation coefficient distribution for Vip3Ag4 at varying concentrations. (**a**) Interference data. Blue 1.0 mg/mL, red 0.5 mg/mL and cyan 0.25 mg/mL; (**b**) Absorbance data. Purple 0.5 mg/mL and yellow 0.25 mg/mL.

2.6. Surface Structure of Vip3Ag

From 100 TEM images of grids incubated with Vip3Ag4, 3492 single particles were picked. Many of these appeared visually to show four-fold symmetry, consistent with our SEC-MALLS and analytical ultracentrifugation (AUC) data (Figure 5). From these particles, EMAN 2 was used to generate a collection of 70 2D class averages (example subset shown in Figure 6a). These were used in the subsequent construction of a modelled surface of the Vip3Ag4 protein (Figure 7a), based on four-fold symmetry as indicated from our SEC-MALLS and SV data. An excellent correlation was observed between 2D class averages and reprojections from the final structure (Figure 6a,b). Fourier shell correlation (FSC) analysis in EMAN2 of structures derived from odd- and even-numbered particles indicated a resolution of approximately 33 Å (FSC = 0.5 criterion: Figure S3). The structure, displayed at a volume threshold consistent with a protein particle of approximately 380 kDa (Figure 7a), shows four clear lobes. When the analysis was repeated in the presence of Ni-NTA-nanogold particles, particle sets that were similar to the original protein were produced (Figure S4a). Reconstruction from the TEM images obtained, based on the initial structure, again gave good correlations between these class averages and the reprojections from the final structure (Figure S4b). Gold particles are clearly visible between the lobes of the tetramer (Figure 7b), indicating binding to the His tags on the recombinant proteins in these locations.

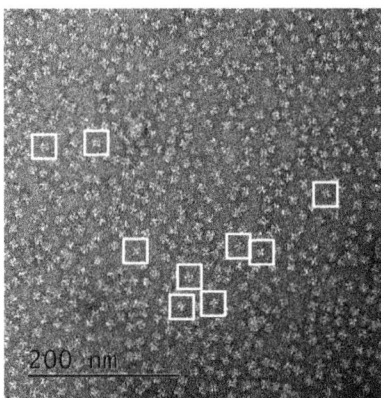

Figure 5. TEM grid. A representative example 40,000× TEM grid is shown with some example single particles captured in 128 × 128 pixel boxes.

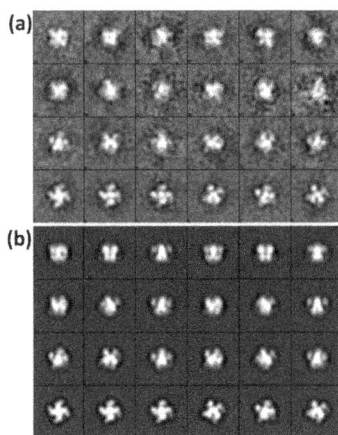

Figure 6. Vip3Ag4 TEM 2D class averages. (**a**) 2D-class averages for Vip3Ag4 particles; (**b**) reprojections from the final 3D model.

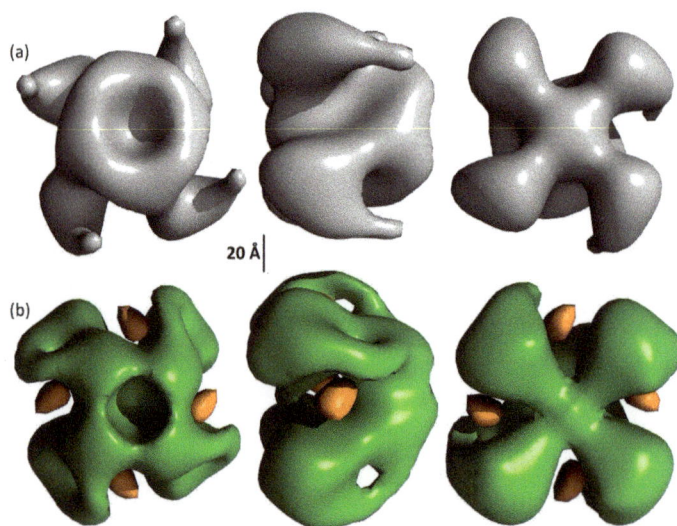

Figure 7. Surface topology of Vip3Ag4. Structures, with (**a**) and without (**b**) nanogold are displayed at volume shells corresponding to the expected molecular mass of Vip3 tetramers (380 kDa). The structure of the protein in the presence of gold is shown in green while the gold is shown in orange. Topology displayed using UCSF Chimera [28].

Analysis of the surface topology of the trypsin-treated Vip3Ag4 again showed good correlation between TEM class averages and reprojections from the model (Figure S5). This structure shows a small region of disconnected density that is likely to be an artifact that may be caused by a bias in the views of the protein on the grid that were used to generate the model. The overall structure shows little change relative to the unprocessed protein (Figure S6) within the resolution of the models. This is consistent with the data above showing that the structure of the complex is maintained with little change following trypsin treatment.

3. Conclusions

These results illustrate for the first time structural data on a member of the Vip3 protein family, at the level of surface topology. Our data indicate that the predominant molecular species of Vip3Ag4 is a tetramer, and that the tetrameric form and general topology are retained after trypsin treatment. Trypsinised Vip3A is able to form pores in artificial bilayers and in *M. sexta* gut cells [12], and pore forming toxins frequently require multimerisation before or during membrane insertion. Whether the tetrameric form of Vip3Ag4 seen here forms a structural precursor to pore formation remains to be established. In addition to providing initial structural insight into the Vip3Ag4 protein, this work establishes that the Vip3Ag4 purity and monodispersity, in addition to the actual size of the tetrameric form, make this protein a candidate for full structural analysis using the emerging technique of cryo-electron microscopy.

4. Materials and Methods

4.1. E. coli Expression and Purification of Vip3Ag4 Protein

The Vip3Ag4 protein was first identified in a Spanish collection of Bt strains and cloned for recombinant expression [29]. Recombinant *E. coli* BL21(DE3) strain harbouring the *vip3Ag4* gene in the pET28b vector was pre-cultured overnight in an orbital shaker at 37 °C and 200 rpm in 2 × YT medium supplemented with 50 µg/mL kanamycin. A 1/25 dilution of this pre-culture into 500 mL 2× YT medium containing 50 µg/mL kanamycin was further incubated for 8 h at 37 °C with vigorous agitation (250 rpm). Protein expression was induced by adding isopropyl-beta-D-1-thiogalactopyranoside to a final concentration of 1 mM and incubation for up to 16 h. Samples were centrifuged at 5000 *g* for 15 min at 4 °C, and the resulting pellet was weighed and resuspended with 3 mL per gram of sonication buffer (20 mM sodium phosphate buffer pH 7.4, 0.5 M NaCl, 3 mg/mL lysozyme (Sigma-Aldrich, Gillingham, United Kingdom), 25 U Benzonase (Novagen, Madison, Wis, USA) and 100 µM phenyl-methylsulfonyl fluoride). Samples were further incubated at 37 °C under gentle agitation for 1 h and were sonicated on ice with a Branson analog sonifier 250 (Branson Ultrasonics Corporation, Danbury, CT, USA) by applying two 1 min pulses with a constant duty cycle at 60 W, separated by a 1 min cooling period. Insoluble material was pelleted at 12,000 *g* for 30 min at 4 °C and the soluble fraction was sequentially filtered through sterile 0.45 and 0.22 µm syringe filters. Protein purification was performed at room temperature (RT) using Protino Ni-TED 2000 Packed Columns following the manufacturer's instructions (Macherey–Nagel, Düren, Germany). After the polyhistidine-tagged protein was eluted, a buffer exchange procedure was performed immediately with Milli-Q water and GE Healthcare PD-10 desalting columns to reduce protein aggregation and precipitation [29]. Subsequently, 5 mL samples were applied to a Hi-Prep 16/60 Sephacryl S-300 size exclusion column equilibrated with 20 mM TrisHCl, pH 8.0, 300 mM NaCl using an AKTA Prime Plus (GE Healthcare, Little Chalfont, UK) at a flow rate of 0.5 mL/min. The same column was also calibrated with gel filtration standards (BioRad, Watford, UK). Fractions (10 mL) were collected and the protein concentration was quantified by the Bradford method [30].

4.2. Trypsin Treatment of Vip3Ag4

Vip3Ag4 was diluted to 0.4 mg/mL in PBS and 200 µL was incubated with 8 µL of 0.1 mg/mL trypsin (a 100:1 *w:w* ratio) at 37 °C for 2 h. The products of the incubation were separated by SDS PAGE (12% acrylamide) with a tricine running buffer. Bands were blotted onto PVDF membrane and revealed by rapid staining with Coomassie blue R-250 and subjected to *N*-terminal sequencing (Abingdon Health Laboratory Services, Birmingham, UK). A sample from the same digestion was also analysed by mass spectrometry (Waters, Elstree, UK, Tri-wave IMS system with integrated high resolution MS capability).

4.3. Mass Spectrometry

Mass spectrometry was carried out using a Waters, Elstree, UK, Synapt G2-Si time of flight mass spectrometer coupled to a Waters Acquity UPLC equipped with a UPLC C4 column held at 60 °C.

4.4. Circular Dichroism

Purified tetrameric Vip3Ag4 protein at 0.25 mg/m was desalted using a PD10 column (GE Healthcare) and analysed using a Chirascan™ CD spectrometer (Applied Photophysics, Leatherhead, UK) at wavelengths between 400–180 nm at 0.5 nm intervals, compared to a sample of its original buffer (20 mM TrisHCl, pH 8.0, 300 mM NaCl) desalted in the same way, as a blank. Data were analysed using the Dichroweb server [31].

4.5. SEC-MALLS

A size exclusion chromatography—multi angle laser light scattering (SEC-MALLS) experiment was performed using a Superdex 200 10/300 Increase column (GE Healthcare) and an AktaPure 25 System (GE Healthcare). The Vip3Ag4 protein sample (100 µL), at a concentration of 1.0 mg/mL, was loaded onto the gel filtration column and eluted with one column volume (24 mL) of 20 mM TrisHCl, 300 mM NaCl buffer, pH 8.0 at a flow rate of 0.7 mL/min. The eluting protein was monitored using a DAWN HELEOS-II 18-angle light scattering detector (Wyatt Technologies, Santa Barbara, CA, USA) equipped with a WyattQELS dynamic light scattering module, a U9-M UV/Vis detector

(GE Healthcare), and an Optilab T-rEX refractive index monitor (Wyatt Technologies). Data were analysed by using Astra (Wyatt Technologies) using the default refractive index increment value of 0.185 mL/g.

4.6. Analytical Ultracentrifugation

Sedimentation velocity (SV) experiments were conducted on a Beckman ProteomeLab XL-I analytical ultracentrifuge using an An-60 Ti rotor at 20 °C. Protein concentrations of 1.0, 0.5 and 0.25 mg/mL, in 20 mM Tris, 300 mM NaCl buffer, pH 8.0, were centrifuged at 20,000 rpm. Absorbance data, at 280 nm, and interference data were collected. Data were analysed using the program SEDFIT, fitting to the c(s) model. The density and viscosity of the buffer were measured using a DMA 5000 M densitometer (Anton-Paar, St Albans, UK) equipped with a Lovis 200 ME viscometer. The protein partial specific volume was calculated using SEDNTERP to allow for the calculation of molecular weight.

4.7. Transmission Electron Microscopy and Single Particle Analysis

Protein samples (100 μg/mL in Tris-HCl pH 8.0, 300 mM NaCl) were adsorbed on to carbon-coated 400 mesh copper grids and negatively stained with 2% (*w/v*) uranyl acetate. Transmission electron microscopy (TEM) images were recorded in the Cardiff University high-resolution TEM facility, using a Jeol JEM-2100 LaB$_6$ transmission electron microscope operating at 200 kV, equipped with a 2 k Gatan Ultrascan camera, at a specimen level increment of 2.47 Å/pixel. EMAN 2 software [32] was used to process 3492 manually selected particles in 128 × 128 pixel boxes. The final model was generated using four-fold symmetry from a total of 14 iterations of refinement, and Fourier shell correlation (FSC) analysis of structures generated from even-and odd-numbered particles indicated a resolution of ~33 Å (FSC = 0.5 criterion).

For gold labelling, purified Vip3Ag4 was incubated for 1 min at room temperature with a 10:1 molar excess of 1.8 nm diameter nickel-nitrilotriacetic acid (Ni-NTA)-Nanogold particles (Nanoprobes, Yaphank, NY, USA). Samples were adsorbed onto grids and washed with Nanogold suspension to prevent unbinding of the protein-associated gold particles. After TEM and image processing, 6286 gold-labeled Vip3Ag4 particles (in 128 × 128 pixel boxes) were automatically selected. Three-dimensional structures were generated using the refined non-labelled Vip3Ag4 structure as a starting model with eight rounds of iterative refinement.

For analysis of the trypsin-treated Vip3Ag4, samples were processed as for untreated protein (above). Following the selection of 3543 particles, EMAN2 was used to construct a new model (without reference to the model for unprocessed protein).

Supplementary Materials: The following are available online at www.mdpi.com/2072-6651/9/5/165/s1, Figure S1: Sequence of the Vip3Ag4 protein. Bold letters represent the sequence added to the N-terminus from the expression construct and including the His-tag. Trypsin cut sites are shown as // above the two amino acids separated by the cleavage. Numbering is given for the full-length, naturally-occurring Vip3Ag4 sequence, Figure S2: CD analysis of Vip3Ag4. The trace for the 0.25 mg/ml Vip3Ag4 sample is shown between 185 and 240 nm. The green curve represents the experimental data, the blue curve the data generated from the reference set, and the pink lines show the difference between the experimental data and the reference data, Figure S3: Vip3Ag4 FSC curve. The curve for the 14th iteration of the EMAN model is shown and indicates a resolution of ~33 Å, Figure S4: Vip3Ag4 nanogold TEM 2D class averages. (a) 2D-class averages for Vip3Ag4 gold-labelled particles. (b) reprojections from the final 3D model, Figure S5: (a) Class averages (left) and 2D-reprojections (right) from the 3D model of trypsin-treated Vip3Ag4. (b) Fourier shell correlation (FSC) of 3D-structures derived from even- and odd-numbered particles indicating a resolution (FSC 0.5 criterion) of approximately 26Å, Figure S6: (a) 3 views of 3D-structure of trypsin-treated Vip3Ag4. A small region of disconnected density (likely to have resulted from some mis-classification of particles in 3D-reconstruction) is indicated with an arrow. (b) Comparison of trypsin-treated (cyan) and native Vip3Ag4 (green) single particle-derived structures, Table S1: Molar mass and hydrodynamic analysis of Vip3Ag4 determined by SEC-MALLS. Table S2: Best-fit parameters obtained from AUC SV data analysis.

Acknowledgments: The authors would like to thank the International Research Support Initiative Program Fellowship, Higher Education Commission of Pakistan for support to SUD. DJS is a Senior Molecular Biology and Neutron Fellow supported by the Science and Technology Facilities Council (UK). GH is funded by the Medical Research Council (UK). Protein purification and TEM work was undertaken in the Cardiff School of Biosciences

Protein Technology Research Hub and Cardiff University High-Resolution TEM facility. Analytical services in the School of Chemistry, Cardiff University, were funded by EPSRC Grant number EP/L027240/1. Open access publishing was funded by Cardiff University.

Author Contributions: L.P., D.J.S., M.T.Y., P.C. and C.B. conceived and designed the experiments; L.P., G.H., S.-U.D., T.L.W., O.J.R. performed the experiments; L.P., D.J.S., G.H., O.R., M.T.Y., P.C. and C.B. analysed the data; L.P., D.J.S., M.T.Y, P.C. and C.B. wrote the paper.

Conflicts of Interest: The authors declare no conflict of interest.

References

1. Van Frankenhuyzen, K. Insecticidal activity of *Bacillus thuringiensis* crystal proteins. *J. Invertebr. Pathol.* **2009**, *101*, 1–16. [CrossRef] [PubMed]

2. Schnepf, E.; Crickmore, N.; Van Rie, J.; Lereclus, D.; Baum, J.; Feitelson, J.; Zeigler, D.R.; Dean, D.H. *Bacillus thuringiensis* and its pesticidal crystal proteins. *Microbiol. Mol. Biol. Rev.* **1998**, *62*, 775–806. [PubMed]

3. Estruch, J.J.; Warren, G.W.; Mullins, M.A.; Nye, G.J.; Craig, J.A.; Koziel, M.G. Vip3A, a novel *Bacillus thuringiensis* vegetative insecticidal protein with a wide spectrum of activities against lepidopteran insects. *Proc. Natl. Acad. Sci. USA* **1996**, *93*, 5389–5394. [CrossRef] [PubMed]

4. Warren, G.W.; Koziel, M.G.; Mullins, M.A.; Nye, G.J.; Carr, B.; Desai, N.M.; Kostichka, K.; Duck, N.B.; Estruch, J.J. Auxiliary proteins for enhancing the insecticidal activity of pesticidal proteins. U.S. Patent 5770696, 23 June 1998.

5. Donovan, W.P.; Engleman, J.T.; Donovan, J.C.; Baum, J.A.; Bunkers, G.J.; Chi, D.J.; Clinton, W.P.; English, L.; Heck, G.R.; Ilagan, O.M.; et al. Discovery and characterization of Sip1A: A novel secreted protein from *Bacillus thuringiensis* with activity against coleopteran larvae. *Appl. Microbiol. Biotechnol.* **2006**, *72*, 713–719. [CrossRef] [PubMed]

6. Crickmore, N.; Baum, J.; Bravo, A.; Lereclus, D.; Narva, K.; Sampson, K.S.; Schnepf, E.; Sun, M.; Zeigler, D.R. *Bacillus thuringiensis* toxin nomenclature. Available online: http://www.btnomenclature.info/ (accessed on 23 November 2016).

7. Sattar, S.; Maiti, M.K. Molecular characterization of a novel vegetative insecticidal protein from *Bacillus thuringiensis* effective against sap-sucking insect pest. *J. Microbiol. Biotechnol.* **2011**, *21*, 937–946. [CrossRef] [PubMed]

8. Barth, H.; Aktories, K.; Popoff, M.R.; Stiles, B.G. Binary bacterial toxins: Biochemistry, biology, and applications of common *Clostridium* and *Bacillus* proteins. *Microbiol. Mol. Biol. Rev.* **2004**, *68*, 373–402. [CrossRef] [PubMed]

9. Han, S.; Craig, J.A.; Putnam, C.D.; Carozzi, N.B.; Tainer, J.A. Evolution and mechanism from structures of an ADP-ribosylating toxin and NAD complex. *Nat. Struct. Biol.* **1999**, *6*, 932–936. [PubMed]

10. Palma, L.; Muñoz, D.; Berry, C.; Murillo, J.; Caballero, P. *Bacillus thuringiensis* toxins: An overview of their biocidal activity. *Toxins* **2014**, *6*, 3296–3325. [CrossRef] [PubMed]

11. Escudero, I.R.; Banyuls, N.; Bel, Y.; Maeztu, M.; Escriche, B.; Munoz, D.; Caballero, P.; Ferre, J. A screening of five *Bacillus thuringiensis* Vip3A proteins for their activity against lepidopteran pests. *J. Invertebr. Pathol.* **2014**, *117*, 51–55. [CrossRef] [PubMed]

12. Lee, M.K.; Walters, F.S.; Hart, H.; Palekar, N.; Chen, J.S. The mode of action of the *Bacillus thuringiensis* vegetative insecticidal protein Vip3A differs from that of Cry1Ab delta-endotoxin. *Appl. Environ. Microbiol.* **2003**, *69*, 4648–4657. [CrossRef] [PubMed]

13. Hamadou-Charfi, D.B.; Boukedi, H.; Abdelkefi-Mesrati, L.; Tounsi, S.; Jaoua, S. *Agrotis segetum* midgut putative receptor of *Bacillus thuringiensis* vegetative insecticidal protein Vip3Aa16 differs from that of Cry1Ac toxin. *J. Invertebr. Pathol.* **2013**, *114*, 139–143. [CrossRef] [PubMed]

14. Abdelkefi-Mesrati, L.; Boukedi, H.; Dammak-Karray, M.; Sellami-Boudawara, T.; Jaoua, S.; Tounsi, S. Study of the *Bacillus thuringiensis* Vip3Aa16 histopathological effects and determination of its putative binding proteins in the midgut of *Spodoptera littoralis*. *J. Invertebr. Pathol.* **2011**, *106*, 250–254. [CrossRef] [PubMed]

15. Sena, J.A.; Hernandez-Rodriguez, C.S.; Ferre, J. Interaction of *Bacillus thuringiensis* Cry1 and Vip3A proteins with *Spodoptera frugiperda* midgut binding sites. *Appl. Environ. Microbiol.* **2009**, *75*, 2236–2237. [CrossRef] [PubMed]

16. Jackson, R.E.; Marcus, M.A.; Gould, F.; Bradley, J.R., Jr.; Van Duyn, J.W. Cross-resistance responses of CrylAc-selected *Heliothis virescens* (Lepidoptera: Noctuidae) to the *Bacillus thuringiensis* protein vip3A. *J. Econ. Entomol.* **2007**, *100*, 180–186. [CrossRef] [PubMed]

17. Chen, Y.; Tian, J.C.; Shen, Z.C.; Peng, Y.F.; Hu, C.; Guo, Y.Y.; Ye, G.Y. Transgenic rice plants expressing a fused protein of Cry1Ab/Vip3H has resistance to rice stem borers under laboratory and field conditions. *J. Econ. Entomol.* **2010**, *103*, 1444–1453. [CrossRef] [PubMed]

18. Chakroun, M.; Banyuls, N.; Bel, Y.; Escriche, B.; Ferre, J. Bacterial Vegetative Insecticidal Proteins (Vip) from Entomopathogenic Bacteria. *Microbiol. Mol. Biol. Rev.* **2016**, *80*, 329–350. [CrossRef] [PubMed]

19. Gomis-Cebolla, J.; Ruiz de Escudero, I.; Vera-Velasco, N.M.; Hernandez-Martinez, P.; Hernandez-Rodriguez, C.S.; Ceballos, T.; Palma, L.; Escriche, B.; Caballero, P.; Ferre, J. Insecticidal spectrum and mode of action of the *Bacillus thuringiensis* Vip3Ca insecticidal protein. *J. Invertebr. Pathol.* **2016**, *142*, 60–67. [CrossRef] [PubMed]

20. Caccia, S.; Chakroun, M.; Vinokurov, K.; Ferre, J. Proteolytic processing of *Bacillus thuringiensis* Vip3A proteins by two *Spodoptera* species. *J. Insect. Physiol.* **2014**, *67*, 76–84. [CrossRef] [PubMed]

21. Abdelkefi-Mesrati, L.; Boukedi, H.; Chakroun, M.; Kamoun, F.; Azzouz, H.; Tounsi, S.; Rouis, S.; Jaoua, S. Investigation of the steps involved in the difference of susceptibility of *Ephestia kuehniella* and *Spodoptera littoralis* to the *Bacillus thuringiensis* Vip3Aa16 toxin. *J. Invertebr. Pathol.* **2011**, *107*, 198–201. [CrossRef] [PubMed]

22. Estruch, J.J.; Yu, C.G.; Warren, G.W.; Desai, N.M.; Koziel, M.G. Plant pest control. Patent WO 9844137, 8 October 1998.

23. Jiang, K.; Mei, S.Q.; Wang, T.T.; Pan, J.H.; Chen, Y.H.; Cai, J. Vip3Aa induces apoptosis in cultured *Spodoptera frugiperda* (Sf9) cells. *Toxicon* **2016**, *120*, 49–56. [CrossRef] [PubMed]

24. Kunthic, T.; Surya, W.; Promdonkoy, B.; Torres, J.; Boonserm, P. Conditions for homogeneous preparation of stable monomeric and oligomeric forms of activated Vip3A toxin from *Bacillus thuringiensis*. *Eur. Biophys. J.* **2016**, *46*, 257–264. [CrossRef] [PubMed]

25. Xiao, Q.; Zhang, F.; Nacev, B.A.; Liu, J.O.; Pei, D. Protein N-terminal processing: Substrate specificity of *Escherichia coli* and human methionine aminopeptidases. *Biochemistry* **2010**, *49*, 5588–5599. [CrossRef] [PubMed]

26. Chakroun, M.; Ferre, J. *In vivo* and *in vitro* binding of Vip3Aa to *Spodoptera frugiperda* midgut and characterization of binding sites by (125)I radiolabeling. *Appl. Environ. Microbiol.* **2014**, *80*, 6258–6265. [CrossRef] [PubMed]

27. Sreerama, N.; Woody, R.W. Estimation of protein secondary structure from circular dichroism spectra: Comparison of CONTIN, SELCON, and CDSSTR methods with an expanded reference set. *Anal. Biochem.* **2000**, *287*, 252–260. [CrossRef] [PubMed]

28. Pettersen, E.F.; Goddard, T.D.; Huang, C.C.; Couch, G.S.; Greenblatt, D.M.; Meng, E.C.; Ferrin, T.E. UCSF Chimera—a visualization system for exploratory research and analysis. *J. Comput. Chem.* **2004**, *25*, 1605–1612. [CrossRef] [PubMed]

29. Palma, L.; de Escuder, I.R.; Maeztu, M.; Caballero, P.; Munoz, D. Screening of *vip* genes from a Spanish *Bacillus thuringiensis* collection and characterization of two Vip3 proteins highly toxic to five lepidopteran crop pests. *Biol. Control* **2013**, *66*, 141–149. [CrossRef]

30. Bradford, M.M. A rapid and sensitive method for the quantitation of microgram quantities of protein utilizing the principle of protein-dye binding. *Anal. Biochem.* **1976**, *72*, 248–254. [CrossRef]

31. Whitmore, L.; Wallace, B.A. Protein secondary structure analyses from circular dichroism spectroscopy: Methods and reference databases. *Biopolymers* **2008**, *89*, 392–400. [CrossRef] [PubMed]

32. Tang, G.; Peng, L.; Baldwin, P.R.; Mann, D.S.; Jiang, W.; Rees, I.; Ludtke, S.J. EMAN2: An extensible image processing suite for electron microscopy. *J. Struct. Biol.* **2007**, *157*, 38–46. [CrossRef] [PubMed]

toxins

MDPI

Article

Listeriolysin O Regulates the Expression of Optineurin, an Autophagy Adaptor That Inhibits the Growth of *Listeria monocytogenes*

Madhu Puri [1], Luigi La Pietra [1], Mobarak Abu Mraheil [1], Rudolf Lucas [2], Trinad Chakraborty [1,*] and Helena Pillich [1,*]

[1] Institute of Medical Microbiology, Justus-Liebig University, 35392 Giessen, Germany; madhu.006.2009@gmail.com (M.P.); luigi.la-pietra@mikrobio.med.uni-giessen.de (L.L.P.); mobarak.mraheil@mikrobio.med.uni-giessen.de (M.A.M.)

[2] Vascular Biology Center, Department of Pharmacology and Toxicology and Division of Pulmonary and Critical Care Medicine, Medical College of Georgia, Augusta University, Augusta, GA 30912, USA; rlucas@augusta.edu

* Correspondence: trinad.chakraborty@mikrobio.med.uni-giessen.de (T.C.); helena.pillich@mikrobio.med.uni-giessen.de (H.P.); Tel.: +49-641-99-41250 (T.C.); +49-641-99-39861 (H.P.)

Academic Editor: Alexey S. Ladokhin
Received: 31 July 2017; Accepted: 2 September 2017; Published: 5 September 2017

Abstract: Autophagy, a well-established defense mechanism, enables the elimination of intracellular pathogens including *Listeria monocytogenes*. Host cell recognition results in ubiquitination of *L. monocytogenes* and interaction with autophagy adaptors p62/SQSTM1 and NDP52, which target bacteria to autophagosomes by binding to microtubule-associated protein 1 light chain 3 (LC3). Although studies have indicated that *L. monocytogenes* induces autophagy, the significance of this process in the infectious cycle and the mechanisms involved remain poorly understood. Here, we examined the role of the autophagy adaptor optineurin (OPTN), the phosphorylation of which by the TANK binding kinase 1 (TBK1) enhances its affinity for LC3 and promotes autophagosomal degradation, during *L. monocytogenes* infection. In LC3- and OPTN-depleted host cells, intracellular replicating *L. monocytogenes* increased, an effect not seen with a mutant lacking the pore-forming toxin listeriolysin O (LLO). LLO induced the production of OPTN. In host cells expressing an inactive TBK1, bacterial replication was also inhibited. Our studies have uncovered an OPTN-dependent pathway in which *L. monocytogenes* uses LLO to restrict bacterial growth. Hence, manipulation of autophagy by *L. monocytogenes*, either through induction or evasion, represents a key event in its intracellular life style and could lead to either cytosolic growth or persistence in intracellular vacuolar structures.

Keywords: listeriolysin O; *Listeria monocytogenes*; optineurin; autophagy

1. Introduction

Listeria monocytogenes is a Gram-positive, ubiquitously distributed, facultative intracellular pathogen that causes listeriosis, a lethal food-borne disease. Following invasion into host cells, the pathogen breaches single-membrane vacuolar compartments to escape into the cytosol using listeriolysin O (LLO) and/or its phospholipases [1,2]. Subsequently, cytosolic bacteria employ the surface protein actin-assembly inducing protein (ActA) to recruit components of the host-cell actin machinery to facilitate intracellular bacterial movement and cell-to-cell spread [1]. However, there is increasing evidence to suggest that a proportion of the bacteria modulate, via LLO, their vacuolar compartments to enable replication and propagation [3,4].

LLO is a cholesterol-dependent cytolysin (CDC) that inserts into host plasma membranes to form pores, thereby inducing host cell signaling cascades that regulate repair processes such as autophagy [5]. LLO is also required for the entry of *L. monocytogenes* into autophagosomal compartments, which fuse with lysosomes, eventually leading to enzymatic degradation [6,7].

Autophagy is a cellular degradation system that involves the enclosure of cargo molecules in double-membrane vacuoles called autophagosomes and their subsequent degradation by lysosomal hydrolases. Autophagic cargo can be comprised of damaged cellular organelles, protein aggregates or pathogens [8]. Autophagy can be triggered by amino acid starvation, low cellular energy levels, withdrawal of growth factors, hypoxia, oxidative stress, endoplasmic reticulum (ER) stress, damaged cellular organelles and infection. Autophagy is an essential part of cellular homeostasis, and an indispensable cellular defense mechanism against intracellular pathogens [8].

Three types of autophagy can occur in cells. Macro-autophagy is the entrapment of cytoplasmic cargo into autophagosomes, followed by fusion with lysosomes leading to subsequent cargo degradation. Micro-autophagy comprises the direct lysosomal uptake of cytosolic components by the invagination of the lysosomal membrane. Chaperone-mediated autophagy involves chaperone proteins that are recognized by the lysosomal membrane receptor lysosome-associated membrane protein 2A. These chaperone proteins form a complex with cargo and are translocated across the lysosomal membrane [9]. Autophagy can also be classified as selective and non-selective. Selective autophagy is mediated by autophagy adaptors or cargo receptors that specifically recognize cargo for degradation, whereas, in non-selective autophagy, cargo is indiscriminately cloistered into developing autophagosomes [10]. Xenophagy is a term used to describe the selective autophagy of intracellular pathogens. In this article, the term "autophagy" refers to the process of xenophagy.

Following induction of autophagy, the cytosol-bound form of microtubule-associated protein 1 light chain 3 (MAP1LC3 or LC3) is converted by the autophagy related proteins (ATG) into its lipidated membrane-bound form, i.e., from LC3-I to LC3-II, by phosphatidylethanolamine conjugation [11]. Autophagy is an innate immune system that restricts the replication of many intracellular pathogens, which include *Salmonella typhimurium*, *Mycobacterium tuberculosis*, *Streptococcus pyogenes* and *Streptococcus pneumoniae* [12–15]. This process requires autophagy adaptors that specifically and selectively recognize intracellular bacteria for their degradation [9]. Autophagy adaptors are characterized by the presence of a ubiquitin-binding domain (UBD) that recognizes ubiquitinated cargo, and an LC3-interacting region (LIR) that links this cargo to the autophagosomal membrane [16]. However, pathogens such as *S. typhimurium* and *L. monocytogenes* have evolved sophisticated strategies to either suppress autophagy, or even prevent recognition and subsequent capture by the autophagic machinery [17–19].

Previous studies have shown that two autophagy adaptors, sequestosome 1 (SQSTM1, also known as p62) and calcium binding and coiled-coil domain 2 (CALCOCO2, also known as NDP52), interact in an exchangeable manner with *L. monocytogenes* to target bacteria to autophagosomal compartments [17,18]. During an examination of the host transcriptional response to LLO, we noted clear and reproducible upregulation of optineurin (OPTN) at very early time points [20]. OPTN has been implicated in many signaling pathways and cellular processes, but, in recent years, its role in autophagy has attracted particular attention. OPTN is now recognized as a member of autophagy adaptors that link LC3 (through their LIR motif) to ubiquitinated cargo (via their UBD) [16]. In the case of *S. typhimurium*, OPTN has been shown to associate with ubiquitinated intracellular bacteria and recruit the TANK binding kinase 1 (TBK1), which enhances OPTN activity [21].

Here, we examined the role of OPTN and the potential interaction between LLO and OPTN in autophagy of *L. monocytogenes*. We show that LLO induces the upregulation of OPTN in HeLa cells. The activation of OPTN was required to restrict the growth of intracellular *L. monocytogenes* wild-type (wt). By contrast, OPTN played no role in the growth restriction of the LLO-negative mutant, the latter of which was able to escape from vacuoles and reach the cytoplasm. Our data demonstrate that OPTN targets *Listeria* to degradation in an LLO-dependent manner.

2. Results

2.1. LC3 Is Essential for the Intracellular Growth Restriction of LLO-Producing L. monocytogenes

In HeLa cells, depletion of the autophagy factor LC3 resulted in a significant increase in intracellular replicating wt *L. monocytogenes* (Figure 1A). These cells are, however, also permissive for the replication of LLO-negative mutants, which exited into the cytoplasm and formed actin tails (Figure 1B). However, depletion of LC3 did not affect the intracellular numbers of LLO-negative *L. monocytogenes* (Figure 1A).

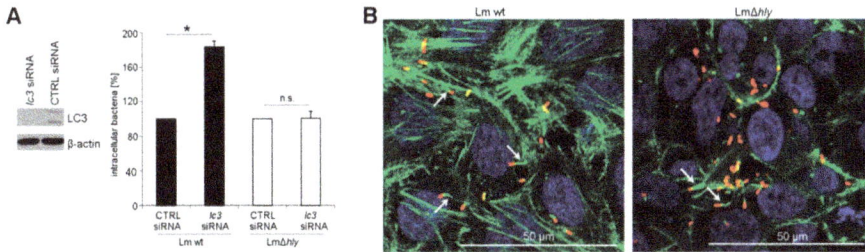

Figure 1. (**A**) Growth of *L. monocytogenes* wt and LLO-negative mutant (Δ*hly*) in LC3-depleted HeLa cells. LC3 depletion was confirmed by immunoblotting of cell lysates, using β-actin as loading control; (**B**) localization of intracellular bacteria as determined by immunofluorescence microscopy. *Listeria* = red, actin cytoskeleton = green, nucleus = blue. Arrows indicate bacterial actin-tails. * $p < 0.05$ vs. CTRL; n.s.: not significant.

2.2. LLO Upregulates OPTN in HeLa Cells

As LLO-producing *L. monocytogenes* were targeted by the autophagy factor LC3, we examined whether LLO regulates OPTN activity. To that purpose, HeLa cells were infected with *L. monocytogenes* wt and LLO-negative mutant. OPTN levels were determined by immunoblotting. As can be seen in Figure 2A, OPTN was upregulated in cells infected with *L. monocytogenes* wt but not *L. monocytogenes* Δ*hly*, indicating that LLO is required to regulate the expression of OPTN. To confirm this result, HeLa cells were treated with lipopolysaccharide (LPS)-free LLO, purified from *L. innocua*, and changes in OPTN expression were analyzed by immunoblotting. Indeed, LLO significantly induced the upregulation of OPTN (Figure 2B).

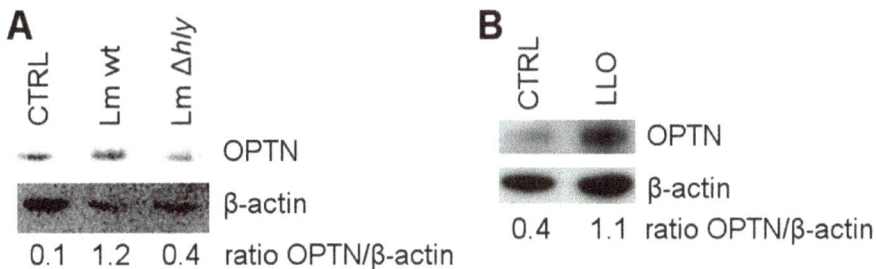

Figure 2. (**A**) Immunoblotting for OPTN in HeLa cells infected with *L. monocytogenes* wt and LLO-negative mutant (Δ*hly*) or left uninfected (CTRL) for 6 h, with β-actin as loading control; (**B**) immunoblotting for OPTN in HeLa cells treated with LLO or left untreated (CTRL), with β-actin as loading control.

2.3. OPTN Phosphorylation by TBK1 Is Essential for the Growth Restriction of L. monocytogenes

Phosphorylation of OPTN by TBK1 enhances its affinity for LC3 [21]. To elaborate the role of TBK1 in *L. monocytogenes* growth restriction, TBK1 was inhibited with a reversible inhibitor BX-795, prior to infection of HeLa cells with wt *L. monocytogenes*. Increased intracellular numbers of wt *L. monocytogenes* were observed in cells treated with BX-795, as compared to untreated control cells (Figure 3A). The treatment of *L. monocytogenes* wt with BX-795 did not affect bacterial viability (Figure S1).

Because TBK1 also phosphorylates other autophagy adaptors, besides OPTN [22], we examined the role of phosphorylated OPTN in *L. monocytogenes* wt growth restriction in greater detail. Cells were co-transfected with a plasmid encoding (1) OPTN and TBK1; or (2) OPTN with TBK1 with an inactive kinase (KM) domain; and (3) an empty vector vehicle. OPTN reduced intracellular wt *L. monocytogenes* growth in the presence of active TBK1, but this effect was absent with the inactive TBK1 variant (Figure 3B). Thus, these results indicate that active TBK1 and phosphorylated OPTN are required to restrict the intracellular growth of *L. monocytogenes*.

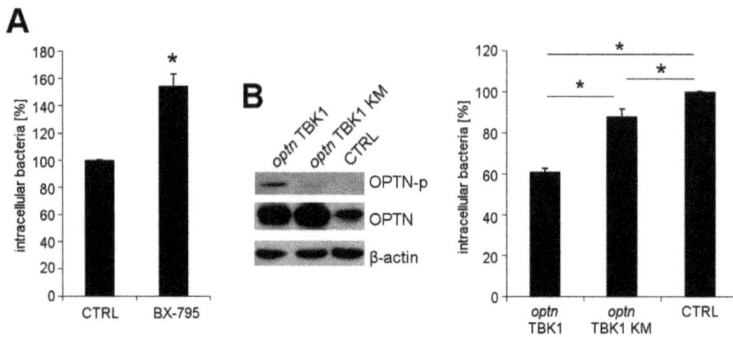

Figure 3. *L. monocytogenes* wt growth in HeLa cells (**A**) treated with BX-795 prior to infection and (**B**) transfected with a plasmid encoding (1) OPTN and TBK1 wt; (2) OPTN and TBK1 with an ineffective kinase (KM); and (3) an empty vector (CTRL). The phosphorylation of OPTN was confirmed by immunoblotting using antibodies against phosphorylated OPTN (OPTN-p), OPTN and β-actin (loading control). * $p < 0.05$ vs. CTRL.

2.4. The Reduction of OPTN Promotes the Growth of Wt L. monocytogenes in an LLO-Dependent Manner

To determine the involvement of LLO in OPTN-mediated growth restriction of *L. monocytogenes*, we reduced expression of *optn* with specific siRNA in HeLa cells, and subsequently infected them with wt *L. monocytogenes* and its isogenic LLO-negative mutant Δ*hly*. In OPTN-depleted cells, the intracellular numbers of wt *L. monocytogenes* were significantly increased. By contrast, OPTN depletion did not affect the intracellular growth of *L. monocytogenes* Δ*hly* (Figure 4). This result implies that LLO production is essential for the growth restriction of *L. monocytogenes* by OPTN.

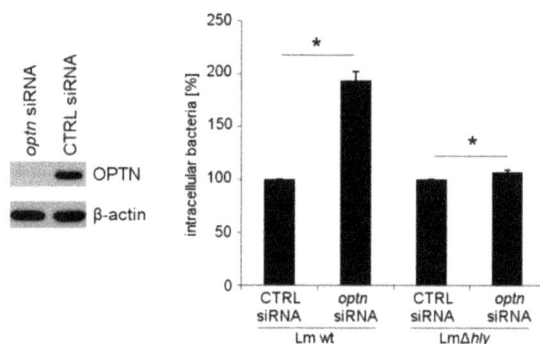

Figure 4. Growth of wt *L. monocytogenes* and LLO-negative mutant (Δ*hly*) in OPTN-depleted HeLa cells. OPTN depletion was confirmed by immunoblotting of cell lysates with β-actin as loading control. * $p < 0.05$ vs. CTRL.

3. Discussion

Autophagy plays a crucial role in the clearance of intracellular *L. monocytogenes* [7,23]. Cytosolic *Listeria* are ubiquitinated and are subsequently detected by the autophagy adaptors SQSTM1 and NDP52, which target them to autophagosomes for degradation [17,18]. Current studies have focused on the question of how *L. monocytogenes* evades autophagic recognition and have provided insight that these bacteria use mimicry, i.e., coating themselves with components of the host cell cytoskeleton by means of ActA [17,24–26]. Our results in this study reveal another aspect of autophagic recognition. Indeed, we report that the autophagy adaptor OPTN is upregulated in response to LLO treatment. Significantly, OPTN reduces the intracellular growth of wt *L. monocytogenes*, but not that of its isogenic LLO-negative mutant strain. Detailed analysis has indicated that TBK1-mediated phosphorylation of OPTN is a crucial event in the restriction of intracellular growth of wt *L. monocytogenes*.

Previous studies on autophagosomal degradation of *L. monocytogenes* have shown that cytoplasmic bacteria are targeted by the autophagosomal machinery [23]. Other reports have demonstrated that LLO is required for autophagy induction, and it was postulated that *L. monocytogenes* containing phagosomes damaged by LLO might be targeted by autophagy [6,7]. The data reported in this study, for the first time, provide evidence that LLO induces the upregulation of the autophagy adaptor OPTN. We used HeLa cells to determine the role of OPTN during *L. monocytogenes* infection. This cell line is particularly well-suited for this study, since expression of LLO is dispensable for bacterial vacuolar escape in these cells [2], as evidenced by the presence of cytoplasmic LLO-negative *L. monocytogenes* with actin tails.

Our data show that intracellular growing *L. monocytogenes* consist of two populations: one which generates LLO and may be targeted by autophagy, thereby leading to its intracellular growth restriction, and a second group that might not be targeted for autophagic clearance and therefore, its growth remains unrestricted. These data therefore suggest that, in addition to evasion of autophagy by ActA [17], *L. monocytogenes* may also manipulate the cellular autophagic machinery by induction through LLO, to promote its growth and persistence in host cells. Thus, bacteria that escape the vacuole and hyper-replicate in the host cytosol may be subjected to autophagic detection and removal (Figure 5). Further studies are required to conclude that autophagy is involved in the growth restriction of LLO-producing *L. monocytogenes* under these experimental conditions. It appears counterintuitive that LLO induces the upregulation of the autophagy adaptor protein OPTN. However, other functions of OPTN may be of relevance here, as it has been shown that the OPTN-TBK1 complex leads to the phosphorylation, dimerization, and nuclear localization of the interferon regulatory factor 3 (IRF3), which, in turn, mediates the transcription of the interferon (IFN) type

1 response genes [27]. Secreted IFNα/β would stimulate the production of more potent antimicrobial interferon IFNγ by bystander cells, subsequently leading to cell-autonomous bacterial killing [28,29]. Thus, our results suggest that LLO induces a host response, the upregulation of OPTN, which is required to detect and to degrade intracellular *L. monocytogenes*.

To date, only one additional bacterial pathogen, namely *S. typhimurium*, was shown to be targeted by OPTN for its autophagosomal degradation [21]. For *Salmonella*, it was demonstrated that LPS leads to TBK1-dependent phosphorylation of OPTN [21], which is a function shared with the proteinaceous toxin LLO. During *S. typhimurium* infection, these bacteria remodel the phagosome into a non-degradative compartment referred to as *Salmonella*-containing vacuole (SCV) [30]. In autophagy-deficient cells, infection with *S. typhimurium* leads to a loss of membrane integrity in SCVs, thus suggesting that autophagy may be involved in membrane repair [31]. There is currently little evidence for repair of host membranes by the autophagic machinery and this certainly requires further investigation.

Our data presented here imply that a quantitative assessment of bacterial replication does not distinguish between the different compartments occupied by the bacterium during intracellular growth. Thus, the compartment in which LLO-deficient bacteria grow in infected cells is not targeted for autophagy and may indeed be the spacious *Listeria*-associated phagosomes previously described, where *L. monocytogenes* grow, albeit at low replication rates [3]. Further studies are warranted to examine replicative niches of *L. monocytogenes* and their contribution to overall growth.

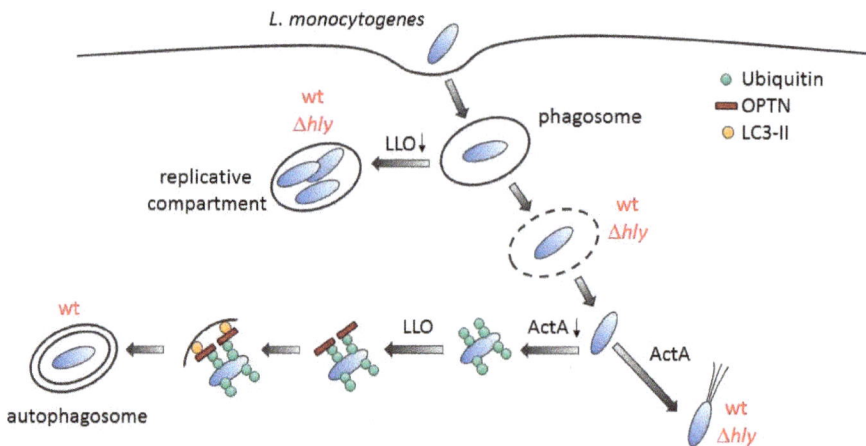

Figure 5. A model for autophagy induction during *L. monocytogenes* infection. After entry, *L. monocytogenes* is trapped within a single-membrane vacuole. Listeriolysin O (LLO)-negative mutant (Δhly) or wild type (wt) bacteria expressing low levels of LLO allow the establishment of a replicative niche, which cannot be autophagocytosed. However, the pathogen escapes from the vacuolar compartment with the help of phospholipases into the cytosol. Cytoplasmic bacteria expressing ActA recruit the host actin cytoskeleton machinery and are camouflaged from autophagic recognition. In contrast, bacteria that do not quickly express ActA are ubiquitinated. This is followed by the binding of ubiquitinated bacteria to OPTN, whose expression is induced by LLO. OPTN interacts with LC3-containing membranes, leading to autophagosome formation around the bacterium.

4. Conclusions

In conclusion, host cells employ OPTN to control the intracellular growth of *L. monocytogenes* via host signaling that is activated by LLO. LLO belongs to the family of CDCs, which are mainly produced by Gram-positive bacteria including species from the genera *Arcanobacterium*, *Bacillus*, *Clostridium*, *Gardnerella*, *Lactobacillus*, *Listeria* and *Streptococcus* [32]. Recently, it was demonstrated that *S. pneumoniae* induces autophagy in a pneumolysin (a CDC)-dependent manner [15]. It might be worth analyzing as to whether or not this toxin activates autophagy via OPTN, as well, which would suggest a general mechanism of CDC-dependent autophagic induction.

5. Materials and Methods

5.1. Cell Culture

HeLa (human cervical adenocarcinoma) cells were cultured in Dulbecco's modified Eagle medium (DMEM) (Thermo Fischer Scientific, Waltham, MA, USA) supplemented with 10% fetal bovine serum (FBS) (Biochrom, Berlin, Germany) at 37 °C in a humidified, 5% CO_2-air atmosphere. The cells were seeded in cell culture dishes with medium containing 10% FBS 24 h prior to the experiments. At 90–100% confluency, the cells were washed once with Hanks' Balanced Salt Solution (HBSS) (Biochrom, Berlin, Germany), and incubated in DMEM containing 10% FBS for 2 h. The cells were then again washed three times with HBSS, and infected in medium containing 0.5% FBS. The cells were incubated in medium containing 0.5% FBS throughout the duration of infection. For treatment with 50 ng/mL LLO, the cells were washed five times with HBSS and incubation with LLO was performed in medium without FBS for 1 h. Prior to treatment, LLO was activated by incubation with 5 mM dithiothreitol (Sigma-Aldrich, St. Louis, MO, USA) for 10 min at room temperature (RT). LLO was isolated and purified from *Listeria innocua* expressing LLO as described [33].

The treatment of cells with 1 µM BX-795 (Merck Millipore, Billerica, MA, USA) was performed 1 h before infection in medium containing 0.5% FBS. The infection was done in the medium containing BX-795.

5.2. RNAi Transfection

The cells were plated shortly before transfection in 1.1 mL DMEM containing 10% FBS. The siRNA (5 nM for *lc3*; 10 nM for *optn*) and the HiPerFect reagent (1.5 µL for *lc3*; 3 µL for *optn*) were diluted in 100 µL DMEM and incubated for 5 min at RT. The transfection complexes were added dropwise to the cells, and the cells were incubated for 48 h. Subsequently, the cells were washed three times with HBSS to terminate the transfection, and DMEM containing 10% FBS was added. The cells were then infected as described. *lc3* (SI02655597), *optn* (SI00132020) and scrambled (1022076) siRNA were purchased from Qiagen (Hilden, Germany).

5.3. Plasmid Transfection

HeLa cells were plated in 24-well plates one day before transfection. Shortly before transfection, the cells were washed five times with sterilized HBSS, incubated in medium without FBS and subsequently transfected with the plasmid pcDNA3.1(+)/HA-OPTN, pcDNA3.1-TBK1-myc-His6, pcDNA3.1-TBK1-myc-His6 KM [21] and the empty vector pRK5 as control (BD Biosciences, Franklin Lakes, NJ, USA). The plasmid DNA (0.95 µg/well) and Lipofectamine 2000 (Invitrogen, Carlsbad, CA, USA; 3 µL/well) were diluted in Opti-MEM I (Thermo Fischer Scientific, Waltham, MA, USA), and equal volumes of both were combined and incubated for 20 min at RT. The plasmid DNA-Lipofectamine 2000 complexes were added to the cells (100 µL/well), and incubated at 37 °C for 4 h. Later, fresh DMEM containing 10% FBS was added, and the cells were infected after 24 h.

5.4. Bacterial Culture and Infection

L. monocytogenes wt (EGD-e) [34] and *L. monocytogenes* Δ*hly* (a mutant lacking LLO) [35] were grown in Brain–Heart–Infusion (BHI) medium. *Escherichia coli* Top 10 (Invitrogen) were cultured in Luria–Bertani medium. The bacteria were grown with constant shaking (180 rpm) at 37 °C. For infection, overnight grown cultures of *L. monocytogenes* were diluted (1:50) in BHI medium, and cultured to exponential growth phase as determined by the optical density at 600 nm. An appropriate culture volume was centrifuged at 13,000 rpm for 1 min at RT. The bacterial pellet was washed twice with HBSS, resuspended in DMEM containing 0.5% FBS, and used for infection. A multiplicity-of-infection of 10 was used for infection. For determination of intracellular bacterial number, the extracellular bacteria were eliminated 1 h post infection (p.i.) by the incubation of the infected cells in DMEM containing 10% FBS, and 50 µg/mL of gentamicin. For analysis of OPTN levels, cells were infected for 6 h without gentamicin treatment.

5.5. Determination of the Number of Intracellular Bacteria

Four hours p.i., the cells were washed three times with phosphate-buffered saline (PBS; pH 7.4), and lysed with cold water containing 0.2% Triton X-100 for 20 min at RT. The bacteria were diluted in PBS and plated on BHI agar plates.

5.6. Protein Preparation from Eukaryotic Cells and Immunoblotting

Cell lysis was performed with RIPA [33] or CHAPS lysis buffer purchased from ProteinSimple (San Jose, CA, USA) [36]. The total protein content was measured with bicinchoninic acid solution (Sigma-Aldrich, St. Louis, MO, USA) assay.

Equal amounts of proteins were analysed by immunoblotting [33]. Antibodies against β-actin (#4970, Cell Signaling Technology, Danvers, MA, USA), LC3 (#sc-16755, Santa Cruz Biotechnology, Dallas, TX, USA), OPTN (#10837-1-AP, Proteintech, Chicago, IL, USA) and phosphorylated-OPTN [21] were used. HRP-conjugated goat anti-rabbit IgG (#sc-2004) and donkey anti-goat IgG (#sc-2020) were purchased from Santa Cruz Biotechnology.

5.7. Immunofluorescence

The cells cultured on coverslips were infected. Four hours p.i., the cells were washed three times with PBS, fixed in 3.7% formaldehyde-PBS for 20 min at RT and incubated with immunofluorescence buffer (0.3% Triton-X-100, 1% BSA in PBS) at RT. After incubation with monoclonal primary anti-*Listeria* antibody (M108, undiluted) overnight at 4 °C, the cells were washed three times with PBS and incubated with 1:1000 anti-mouse IgG Fab2 Fragment Alexa Fluor 647-conjugated secondary antibody (Cell Signaling Technology, Danvers, MA, USA, #4410) and 1:40 Alexa Fluor 488-conjugated phalloidin (Thermo Fisher Scientific, Waltham, MA, USA, #A12379) for 2 h at 37 °C in the dark. After three washing steps, the coverslips were mounted with ProLong Gold antifade reagent with DAPI (Thermo Fisher Scientific, Waltham, MA, USA, #P36935) and imaged by confocal microscopy (Leica TCS SP5, Leica Microsystems, Wetzlar, Germany).

5.8. Statistical Analysis

Statistical analysis of experiments was performed with SigmaPlot 11 (Systat Software, San Jose, CA, USA). The data of Figures 1A, 3A, 4 and S1 were analyzed by *t*-test. The data of Figure 3B were analyzed by one-way ANOVA with Tukey. Mean values ± SEM are plotted from three independent experiments. Representative immunofluorescence or immunoblotting images from three independent experiments are shown.

Supplementary Materials: The following are available online at www.mdpi.com/2072-6651/9/9/273/s1, Figure S1: BX-795 has no effect on bacterial viability. *L. monocytogenes* wt was added to the cell culture media (without cells) containing 1 µM BX-795. The bacteria were plated after 1 h.

Acknowledgments: We thank Sylvia Krämer, Nelli Schklarenko and Natalia Lest for technical support, and Ivan Dikic for providing the anti-phosphorylated-OPTN antibody and OPTN and TBK1 plasmids. This study was supported by the Bundesministerium für Bildung und Forschung (BMBF) ERA-NET PathoGenoMics LISTRESS and PROANTILIS, to T.C., as well as by an Extramural Success Award from the Vice President for Research at Augusta University (to R.L.) and the SFB grant TR-84 "Innate Immunity of the Lung" from the German Research Foundation (DFG) (to H.P. and T.C., project A04). R.L. is a Mercator Fellow of the DFG.

Author Contributions: T.C. and H.P. conceived and designed the experiments; M.P., L.L.P. and H.P. performed the experiments; M.P., M.A.M., R.L., T.C. and H.P. analyzed the data; M.P., R.L., T.C. and H.P. wrote the paper.

Conflicts of Interest: The authors declare no conflict of interest.

References

1. Disson, O.; Lecuit, M. In vitro and in vivo models to study human listeriosis: Mind the gap. *Microbes Infect.* **2013**, *15*, 971–980. [CrossRef] [PubMed]
2. Gründling, A.; Gonzalez, M.D.; Higgins, D.E. Requirement of the *Listeria monocytogenes* broad-range phospholipase PC-PLC during infection of human epithelial cells. *J. Bacteriol.* **2003**, *185*, 6295–6307. [CrossRef] [PubMed]
3. Birmingham, C.L.; Canadien, V.; Kaniuk, N.A.; Steinberg, B.E.; Higgins, D.E.; Brumell, J.H. Listeriolysin O allows *Listeria monocytogenes* replication in macrophage vacuoles. *Nature* **2008**, *451*, 350–354. [CrossRef] [PubMed]
4. Lam, G.Y.; Cemma, M.; Muise, A.M.; Higgins, D.E.; Brumell, J.H. Host and bacterial factors that regulate LC3 recruitment to *Listeria monocytogenes* during the early stages of macrophage infection. *Autophagy* **2013**, *9*, 985–995. [CrossRef] [PubMed]
5. Meyer-Morse, N.; Robbins, J.R.; Rae, C.S.; Mochegova, S.N.; Swanson, M.S.; Zhao, Z.; Virgin, H.W.; Portnoy, D. Listeriolysin O is necessary and sufficient to induce autophagy during *Listeria monocytogenes* infection. *PLoS ONE* **2010**, *5*, e8610. [CrossRef] [PubMed]
6. Birmingham, C.L.; Canadien, V.; Gouin, E.; Troy, E.B.; Yoshimori, T.; Cossart, P.; Higgins, D.E.; Brumell, J.H. *Listeria monocytogenes* evades killing by autophagy during colonization of host cells. *Autophagy* **2007**, *3*, 442–451. [CrossRef] [PubMed]
7. Py, B.F.; Lipinski, M.M.; Yuan, J. Autophagy limits *Listeria monocytogenes* intracellular growth in the early phase of primary infection. *Autophagy* **2007**, *3*, 117–125. [CrossRef] [PubMed]
8. Lin, L.; Baehrecke, E.H. Autophagy, cell death, and cancer. *Mol. Cell Oncol.* **2015**, *2*, e985913. [CrossRef] [PubMed]
9. Glick, D.; Barth, S.; Macleod, K.F. Autophagy: Cellular and molecular mechanisms. *J. Pathol.* **2010**, *221*, 3–12. [CrossRef] [PubMed]
10. Moy, R.H.; Cherry, S. Antimicrobial autophagy: A conserved innate immune response in Drosophila. *J. Innate Immun.* **2013**, *5*, 444–455. [CrossRef] [PubMed]
11. Bento, C.F.; Renna, M.; Ghislat, G.; Puri, C.; Ashkenazi, A.; Vicinanza, M.; Menzies, F.M.; Rubinsztein, D.C. Mammalian autophagy: How does it work? *Annu. Rev. Biochem.* **2016**, *85*, 685–713. [CrossRef] [PubMed]
12. Zheng, Y.T.; Shahnazari, S.; Brech, A.; Lamark, T.; Johansen, T.; Brumell, J.H. The adaptor protein p62/SQSTM1 targets invading bacteria to the autophagy pathway. *J. Immunol.* **2009**, *183*, 5909–5916. [CrossRef] [PubMed]
13. Gutierrez, M.G.; Master, S.S.; Singh, S.B.; Taylor, G.A.; Colombo, M.I.; Deretic, V. Autophagy is a defense mechanism inhibiting BCG and *Mycobacterium tuberculosis* survival in infected macrophages. *Cell* **2004**, *119*, 753–766. [CrossRef] [PubMed]
14. Nakagawa, I.; Amano, A.; Mizushima, N.; Yamamoto, A.; Yamaguchi, H.; Kamimoto, T.; Nara, A.; Funao, J.; Nakata, M.; Tsuda, K.; et al. Autophagy defends cells against invading group A Streptococcus. *Science* **2004**, *306*, 1037–1040. [CrossRef] [PubMed]
15. Li, P.; Shi, J.; He, Q.; Hu, Q.; Wang, Y.Y.; Zhang, L.J.; Chan, W.T.; Chen, W.X. *Streptococcus pneumoniae* induces autophagy through the inhibition of the PI3K-I/Akt/mTOR pathway and ROS hypergeneration in A549 cells. *PLoS ONE* **2015**, *10*, e0122753. [CrossRef] [PubMed]
16. Stolz, A.; Ernst, A.; Dikic, I. Cargo recognition and trafficking in selective autophagy. *Nat. Cell Biol.* **2014**, *16*, 495–501. [CrossRef] [PubMed]
17. Yoshikawa, Y.; Ogawa, M.; Hain, T.; Yoshida, M.; Fukumatsu, M.; Kim, M.; Mimuro, H.; Nakagawa, I.; Yanagawa, T.; Ishii, T.; et al. *Listeria monocytogenes* ActA-mediated escape from autophagic recognition. *Nat. Cell Biol.* **2009**, *11*, 1233–1240. [CrossRef] [PubMed]
18. Mostowy, S.; Sancho-Shimizu, V.; Hamon, M.A.; Simeone, R.; Brosch, R.; Johansen, T.; Cossart, P. p62 and NDP52 proteins target intracytosolic *Shigella* and *Listeria* to different autophagy pathways. *J. Biol. Chem.* **2011**, *286*, 26987–26995. [CrossRef] [PubMed]

19. Thomas, M.; Mesquita, F.S.; Holden, D.W. The DUB-ious lack of ALIS in Salmonella infection: A Salmonella deubiquitinase regulates the autophagy of protein aggregates. *Autophagy* **2012**, *8*, 1824–1826. [CrossRef] [PubMed]

20. Ghai, R. Transcriptional Response of Murine Bone Marrow Macrophages to Listeriolysin, the Pore-Forming Toxin of *Listeria monocytogenes*. Ph.D. Thesis, Justus Liebig University, Giessen, Germany, 2006.

21. Wild, P.; Farhan, H.; McEwan, D.G.; Wagner, S.; Rogov, V.V.; Brady, N.R.; Richter, B.; Korac, J.; Waidmann, O.; Choudhary, C.; Dötsch, V.; et al. Phosphorylation of the autophagy receptor optineurin restricts *Salmonella* growth. *Science* **2011**, *333*, 228–233. [CrossRef] [PubMed]

22. Pilli, M.; Arko-Mensah, J.; Ponpuak, M.; Roberts, E.; Master, S.; Mandell, M.A.; Dupont, N.; Ornatowski, W.; Jiang, S.; Bradfute, S.B.; et al. TBK-1 promotes autophagy-mediated antimicrobial defense by controlling autophagosome maturation. *Immunity* **2012**, *37*, 223–234. [CrossRef] [PubMed]

23. Rich, K.A.; Burkett, C.; Webster, P. Cytoplasmic bacteria can be targets for autophagy. *Cell Microbiol.* **2003**, *5*, 455–468. [CrossRef] [PubMed]

24. Dortet, L.; Mostowy, S.; Samba-Louaka, A.; Gouin, E.; Nahori, M.A.; Wiemer, E.A.; Dussurget, O.; Cossart, P. Recruitment of the major vault protein by InlK: A *Listeria monocytogenes* strategy to avoid autophagy. *PLoS Pathog.* **2011**, *7*, e1002168. [CrossRef]

25. Mitchell, G.; Ge, L.; Huang, Q.; Chen, C.; Kianian, S.; Roberts, M.F.; Schekman, R.; Portnoy, D.A. Avoidance of autophagy mediated by PlcA or ActA is required for *Listeria monocytogenes* growth in macrophages. *Infect. Immun.* **2015**, *83*, 2175–2184. [CrossRef] [PubMed]

26. Pillich, H.; Puri, M.; Chakraborty, T. ActA of *Listeria monocytogenes* and Its Manifold Activities as an Important Listerial Virulence Factor. *Curr. Top. Microbiol. Immunol.* **2017**, *399*, 113–132. [PubMed]

27. Bakshi, S.; Taylor, J.; Strickson, S.; McCartney, T.; Cohen, P. Identification of TBK1 complexes required for the phosphorylation of IRF3 and the production of interferon β. *Biochem. J.* **2017**, *474*, 1163–1174. [CrossRef] [PubMed]

28. Alsharifi, M.; Müllbacher, A.; Regner, M. Interferon type I responses in primary and secondary infections. *Immunol. Cell Biol.* **2008**, *86*, 239–245. [CrossRef] [PubMed]

29. Jin, Y.; Lundkvist, G.; Dons, L.; Kristensson, K.; Rottenberg, M.E. Interferon-gamma mediates neuronal killing of intracellular bacteria. *Scand. J. Immunol.* **2004**, *60*, 437–448. [CrossRef] [PubMed]

30. Bakowski, M.A.; Braun, V.; Brumell, J.H. Salmonella-containing vacuoles: Directing traffic and nesting to grow. *Traffic* **2008**, *9*, 2022–2031. [CrossRef] [PubMed]

31. Kreibich, S.; Emmenlauer, M.; Fredlund, J.; Rämö, P.; Münz, C.; Dehio, C.; Enninga, J.; Hardt, W.D. Autophagy proteins promote repair of endosomal membranes damaged by the Salmonella type three secretion system 1. *Cell Host Microbe* **2015**, *18*, 527–537. [CrossRef] [PubMed]

32. Hotze, E.M.; Tweten, R.K. Membrane assembly of the cholesterol-dependent cytolysin pore complex. *Biochim. Biophys. Acta* **2012**, *1818*, 1028–1038. [CrossRef] [PubMed]

33. Pillich, H.; Loose, M.; Zimmer, K.P.; Chakraborty, T. Activation of the unfolded protein response by *Listeria monocytogenes*. *Cell Microbiol.* **2012**, *14*, 949–964. [CrossRef] [PubMed]

34. Glaser, P.; Frangeul, L.; Buchrieser, C.; Rusniok, C.; Amend, A.; Baquero, F.; Berche, P.; Bloecker, H.; Brandt, P.; Chakraborty, T.; et al. Comparative genomics of *Listeria* species. *Science* **2001**, *294*, 849–852. [PubMed]

35. Guzman, C.A.; Rohde, M.; Chakraborty, T.; Domann, E.; Hudel, M.; Wehland, J.; Timmis, K.N. Interaction of *Listeria monocytogenes* with mouse dendritic cells. *Infect. Immun.* **1995**, *63*, 3665–3673. [PubMed]

36. Loose, M.; Hudel, M.; Zimmer, K.P.; Garcia, E.; Hammerschmidt, S.; Lucas, R.; Chakraborty, T.; Pillich, H. Pneumococcal hydrogen peroxide-induced stress signaling regulates inflammatory genes. *J. Infect. Dis.* **2015**, *211*, 306–316. [CrossRef] [PubMed]

MDPI AG
St. Alban-Anlage 66
4052 Basel, Switzerland
Tel. +41 61 683 77 34
Fax +41 61 302 89 18
http://www.mdpi.com

Toxins Editorial Office
E-mail: toxins@mdpi.com
http://www.mdpi.com/journal/toxins

www.ingramcontent.com/pod-product-compliance
Lightning Source LLC
Chambersburg PA
CBHW051912210326
41597CB00033B/6116